1 MONTH OF
FREE
READING

at

www.ForgottenBooks.com

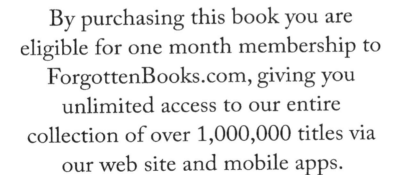

By purchasing this book you are eligible for one month membership to ForgottenBooks.com, giving you unlimited access to our entire collection of over 1,000,000 titles via our web site and mobile apps.

To claim your free month visit:
www.forgottenbooks.com/free1233960

ISBN 978-0-332-73009-7

PIBN 11233960

PRESENTED TO THE
NATIONAL PARK SERVICE
SOUTHEAST REGION

FRASER FIR IN THE GREAT SMOKY MOUNTAINS NATIONAL PARK

ITS DEMISE BY THE BALSAM WOOLLY APHID (ADELGES PICEAE RATZ.)

BY

RONALD L. HAY
C. CHRISTOPHER EAGAR
KRISTINE D. JOHNSON

THE DEPARTMENT OF FORESTRY, WILDLIFE AND FISHERIES
THE UNIVERSITY OF TENNESSEE

PREFACE

The magnificent spruce-fir forests in the Great Smoky Mountains have long been a source of awe and inspiration to all people. Legends steeped in mystery tell of great birds and other creatures that roamed these heights creating voids in the dense, evergreen forests with their footprints. Panoramic views from Mt. LeConte, the Jumpoff, and Mt. Guyot have penetrated to the inner depth of human souls since they were first witnessed, and they continue to have a major impact on visitors today.

During the pre-Park period, these spruce-fir forests were sporadically logged and some settlers grazed livestock in the scattered grassy balds, but these mountain heights continued to supply unceasing inspiration to those hardy enough to seek their summits. The desire to share such experiences was part of the motivation that lead to founding Mt. LeConte Lodge and the Great Smoky Mountains National Park. Since that time the sanctity of these mountains has been protected from most disturbances, certainly all that would have irreparable consequences.

Millions of visitors have shared the spruce-fir forests in the Great Smoky Mountains National Park from the Transmountain Road; fewer have explored isolated areas along the Appalachian Trail and still fewer of us have witnessed the most pristine wilderness available in the East by leaving the trails, BUT we all have gained from our experiences. Now, however, a Park visitor is proving to be most unwelcome. Balsam woolly aphids came to the Great Smoky Mountains National Park prior to 1958 and they have not left!

Those attributes of the spruce-fir forests that were once revered by so many human visitors are now being catastrophically defiled by these insects. Fraser fir trees are being killed by the millions and the sanctity of these magnificent forests is being destroyed at the same time. The summit of Mt. Sterling is already an accumulation of dead fir; more peaks will surely reaffirm the destructive impact of this insect.

There is no acceptable chemical control for the insect and biological controls have not yet been effective. In the face of such a verdict, this project was undertaken to study pertinent characteristics of aphid attack and subsequent affects on the fir community. It is our fervid hope that a meaningful understanding of the future of Fraser fir in the Southern Appalachians will come from our work. There are yet generations of man that need these mountains.

Many people and agencies have provided assistance during all phases of this study. The National Park Service provided funding and guidance for the study. We greatly appreciated and enjoyed the total support of the Southeast Region, Dr. Raymond Herrmann, Chief Scientist; the Great Smoky Mountains National Park, Mr. Boyd Evison, Superintendent; and the Uplands Field Research Laboratory, Dr. Susan P. Bratton, Director.

Aerial, color infrared transparencies, plus unlimited council were provided by the Forest Insect and Disease Management Division, State and Private Forestry, U. S. Forest Service, Harold Flake,

Field Representative. Special appreciation is due Dr. William H.
Sites, U. S. Forest Service because he is a friend and for sugges-
tions and unceasing help during the planning of the project photography.

The Agricultural Experiment Station, University of Tennessee,
Dr. D. M. Gossett, Dean, supported and coordinated the logistics
for the project. We appreciate their willingness to sponsor and
cooperate with such research projects.

Special appreciation must be expressed to C. Christopher
Eagar and Kristine Diane Johnson, two friends who worked so many
long hours on all aspects of this project. Truly none of this
could have been accomplished with excellence without them. What
a pleasure to share a campfire and a project with each of them!

Ronald L. Hay

Department of Forestry, Wildlife and Fisheries
The University of Tennessee
Knoxville

January 1, 1978

CONTENTS

LIST OF TABLES

LIST OF FIGURES

FRASER FIR IN THE GREAT SMOKY
MOUNTAINS NATIONAL PARK

CONCLUSIONS

BALSAM WOOLLY APHIDS HAVE COLONIZED THE ENTIRE SPRUCE-FIR TYPE IN THE GREAT SMOKY MOUNTAINS NATIONAL PARK.

Fir mortality was greatest in the northeastern areas of the Park, but many Fraser fir were dying on all prominent mountains. Infestations were above 1950 meters on Mt. Guyot and Mt. LeConte and near 1829 meters on Clingmans Dome. Although infestation development is somewhat dependant on the environment, aphid populations should consume the fir on Mt. Guyot in 3 to 5 years, on Mt. LeConte in 5 to 10 years, and on Clingmans Dome in 10 to 15 years.

BALSAM WOOLLY APHIDS PREFERRED COMMUNITIES WITH MATURE FRASER FIR AND RED SPRUCE IN WHICH TO INITIATE INFESTATIONS.

Random, passive dissemination of aphids probably placed them into many communities, but those with maturing Fraser fir were most efficient in intercepting aphids and the large fir provided adequate attack sites plus abundant energy for developing large aphid populations. Dense, pure, even-aged Fraser fir stands at high elevations were nearly free of aphids. However, there was limited reason to assume that they would remain healthy.

BALSAM WOOLLY APHID INFESTATIONS APPEARED AT THE NORTHERN HARDWOOD-
FIR ECOTONE AND SPREAD TOWARD THE SUMMIT WITH INCREASING SPEED AS
THE APHID POPULATION EXPANDED.

Initial infestations found along the northern
hardwood-fir ecotones were small, numerous, and scattered. As they expanded and merged, large infestations
grew and provided aphids which convective winds moved
upslope. Future analyses will indicate the speed of
the aphid population development along elevation gradients.

PERMANENT PLOTS WERE ESTABLISHED THROUGHOUT THE PROJECT.

A system of permanent plots was established
throughout the spruce-fir type in the Great Smoky
Mountains National Park. They were marked in the
field, on aerial transparencies, on maps, and they
were filed with National Park Service, Southeast
Region, Chief Scientist, Atlanta. Future use of
these plots might include more than balsam woolly
aphid studies.

IN ABSENCE OF EFFECTIVE BALSAM WOOLLY APHID CONTROLS APPLICABLE
WITHIN WILDERNESS AREAS, MANAGEMENT RECOMMENDATIONS WERE SUGGESTED.

Management recommendations included provisions to minimize wildfire risk and hazard by mani-
pulating the people-visits and the fuels, provisions
to encourage and secure Fraser fir reproduction in
selected stands, and provisions to perpetuate an "endangered species" through a cooperative seed orchard
program.

CHAPTER I

THE BALSAM WOOLLY APHID COMES
TO THE SOUTHERN APPALACHIANS

The balsam woolly aphid (Adelges piceae (Ratz.) belongs to
the insect order Homoptera (Hemiptera), family Phylloxeridae
(Chermidae, Psyllidae, or jumping plant-lice), super family
Aphidoidea and sub-family Adelginae. The adelgids infest only
conifers. Of the eleven species known to attack the true firs
(Abies), all are holarctic in origin and they appear to have
evolved into an ecological balance with their native hosts,
becoming problematic only when they are introduced outside their
natural range.

Adelges piceae (Ratz.) originated in Eurpoe, probably as a
derivative of the closely related A. nusslini (Borner), where
the latter was a relatively innoculous pest on European silver
fir (Abies alba (L.) Mill). Its history on the North American
continent has been similar to other disasterous plant pests;
introduced accidentally on nursery stock, the woolly aphid thrived
and spread so rapidly that it was already well extablished on balsam fir (Abies balsamea (L.) Mill) in Maine and Maritime Canada
before official identification was made in 1908 (Kotinsky, 1916).
Damage to this commercially important fir has been extensive and

occasionally severe throughout the Northeast.

A separate introduction from Europe put the woolly aphid into the Western United States, where it first appeared around 1928 on noble fir (Abies procera Rehd.) and grand fir (Abies grandis (Dougl.) Lindl.) near San Francisco (Annand, 1928). Damage is now widespread in the Pacific Northwest on Pacific silver fir (Abies amabilis (Dougl.) Fordes) and subalpine fir (Abies lasiocarpa (Hook.) Nutt.). With the possible exception of Abies bracteata (D. Don) all North American fir have been attacked by the balsam woolly aphid. Some species show more susceptibility to the insect than others; balsam and Fraser fir (Abies fraseri (Pursh) Poir.) have high suscepti- bility with mortality likely, grand fir and shasta red fir (Abies magnifica var. shastensis Lemm.) have moderate susceptibility with only occasional mortality, while noble fir and white fir (Abies concolor (Gord. and Glend.) Lindl.) have low susceptibility with slight mortality (Bryant, 1974).

Economic damage caused by the balsam woolly aphid in North America has been extensive, and for some species it has been locally heavy. Reed (1964) reported that grand and Pacific silver fir were killed on 242,812 hectares in Oregon and Washington. Bryant (1974) found 284,899 square kilometers of balsam fir type damaged by the aphid through 1968. Damage to the timber resource has not been fre- quently assessed, however. Perhaps, the lack of an effective control, either biological, managerial, or chemical, so totally pointed to

the frustrations of forest managers that no one really wanted to know exactly what was being lost. There has been renewed interest within the profession concerning balsam woolly aphid control problems.

In those areas where aphids have been resident for many years, timber damage may have become tolerable. For example, the balsam woolly aphid was introduced about 1890 into northeastern North America and subsequently infested vast areas, yet balsam fir remains an important species, at least in part due to some environmental controls. Although -33.8 C for one night did not kill aphids on Mt. Mitchell, N. C. (Amman, 1967), balsam fir has resumed near normal growth after the aphid was controlled. Occasionally Fraser fir saplings have also shown this characteristic. Some fir species may warrant endangered species status due to balsam woolly aphid attack, e. g., subalpine fir and Fraser fir.

Both Fraser fir and subalpine fir have disjunct ranges within which the aphid could isolate and eradicate the fir hosts. This is critical for Fraser fir which holds to a remnant distribution on a few isolated mountain peaks in the Southern Appalachians that approximate 1829 meters elevation. Since the introduction of the aphid into this region, Fraser fir has been removed from the overstory on most mountain peaks.

Balsam Woolly Aphids Move into the Southern Appalachians

The balsam woolly aphid was first discovered in the Southern Appalachian Mountains on Mt. Mitchell below Stepp Gap in 1957.

Discovery of the aphid was the result of an investigation initiated
by the District Ranger on Toecanoe District of the Pisgah National
Forest; there were several groups of from 10 to 200 dead fir trees
on his district in 1955. Subsequent investigations confirmed
the presence of the balsam woolly aphid (Speers, 1958). Expanding
reconnaissance in 1958 revealed 11,000 dead fir trees on Mt. Mitchell
and all 3,035 hectares of resident Fraser fir type were infested.
By 1963, 275,000 Fraser firs were dead on Mt. Mitchell. Current
surviving fir trees in the upper canopy were restricted to those
areas sprayed with benzene hexochloride at frequent intervals.

Balsam woolly aphids had apparently infested the fir stands on
Mt. Mitchell for many years prior to their discovery. It is known
that 11,000 trees were already dead in several different areas when
the infestation was first acknowledged. Such widespread dispersal
of the pest throughout the host distribution is strong evidence of
its long history on Mt. Mitchell. Although aphid populations build
extremely fast, it still takes thousands of them from two to seven
years to kill host trees. Speculatively back-dating from the confir-
mation of balsam woolly aphid presence, places its arrival on Mt.
Mitchell as late as the 1940's, but more likely the introduction occurred during the mid- to late-1930's. Population expansion between
then and 1958 could readily account for 11,000 dead trees on 3,035
infested hectares. One 7 to 10 year cycle would not have been adequate to wreck such total havoc on so many trees.

Distribution of fir species is not contiguous from balsam woolly aphid infestations in the Northeast into the Southern Appalachians. Balsam and Fraser firs are separated in distribution by an intermediate form, Abies balsamea var. phanerolepis (Fernald) which occurs as disjunct communities along the Blue Ridge Mountains in Virginia and West Virginia. Aphids had infested these stands for some time and were identified in 1956 (McCambridge and Kowal, 1957) before they first appeared on Mt. Mitchell. How then did the balsam woolly aphid make the big jump from the infestations in Virginia to Mt. Mitchell in North Carolina?

The invasion. Wind is a possibility. Albeit wingless, aphids are light enough that winds push them for considerable distances and altitudes. Winds are suspected as being the most prevalent dispersal mechanism, but the maximum distance of wind dissemination is not known. However, it is possible aphids could have leap-frog from one fir community to the next along the crest of the Blue Ridge into North Carolina.

Animal vectors were also a possibility. Birds are known to disperse aphids within a stand during feeding. People were suspected of being balsam woolly aphids vectors. Amman (1966) postulated that woodcutters working on private land near Mt. Mitchell may have been responsible for transmitting the aphid from Mt. Mitchell from Roan Mountain, Tennessee; they moved from a timber sale on Mt. Mitchell to a Roan Mountain sale. Recreationists may transmit aphids, particularly hikers or campers in recreational vehicles visiting several

campgrounds. Personal communications with the Forest Insect and Disease Management Division of the U. S. Forest Service, Asheville, North Carolina, has diminished the likelihood of human vectors, however. During an extensive survey in the Great Smoky Mountains, thorough personnel checks did not reveal living aphids on their clothing.

Extensive use of many fir species as ornamentals may have facilitated the distribution of woolly aphids. _Abies_ are desirable for landscape plantings and they enjoy extensive distribution through-out the mountains in the East. Aphids which are difficult to de-tect on small trees could have reached the Southern Appalachians on fir ornamentals.

Speculative timing of the initial infestation on Mt. Mitchell adds credence to reforestation programs as a possible vector. A major objective of the Civilian Conservation Corps (CCC) was to restore abused and depleted forest resources. Millions of seed-lings were planted throughout the United States during the 1930's by CCC and similar programs. The demand for planting stock was so strong that pest problems were not seriously considered. The main prerequisites for seedlings were leaves on top and roots to put in the ground.

Clearly the spruce-fir forests in the Southern Appalachians were a depleted resource. Extensive logging without prescriptions for stand regeneration had rapidly capped exploitation of these for-ests. The original spruce-fir forests covered 607,029 hectares.

9

By 1860 only 303,514 hectares remained, and subsequent exploitation
further reduced the spruce-fir to 20,234 hectares by 1960. If fir
trees were planted in North Carolina during the 1930's, it's possible that once again seedlings were a vector for balsam woolly
aphids.

The Aphid Project in the Great Smokies

Mt. Sterling in the Great Smoky Mountains National Park is
64.4 kilometers, S 85° W of Mt. Mitchell (see Figure 1-1). During aerial reconnaissance for balsam woolly aphids in 1963, an
infestation was found on Mt. Sterling Ridge just north of the
summit. There were 45 dead trees in an infested area of 6 to 12
hectares.

The largest dying fir had a 21.6 centimeter diameter with
sufficient "rotholz" (Doerkson and Mitchell, 1965) to show aphid
attack in 1958. However, without clear evidence as to when the
other dead trees had died, it was impossible to accurately date
the aphid invasion of the Park (Amman, 1966).

Winds were probably responsible for aphids reaching the Park.
U. S. Weather Bureau records showed that east winds with velocities
exceeding 56 kilometers per hour for up to 3 days were not uncom-
mon during those months of suspected aphid arrival on Mt. Sterling.
The infested area on Mt. Sterling was not readily accessible to
human visitors, thereby decreasing the chance of human vector.

The Mt. Sterling infestation was the first presence of

10

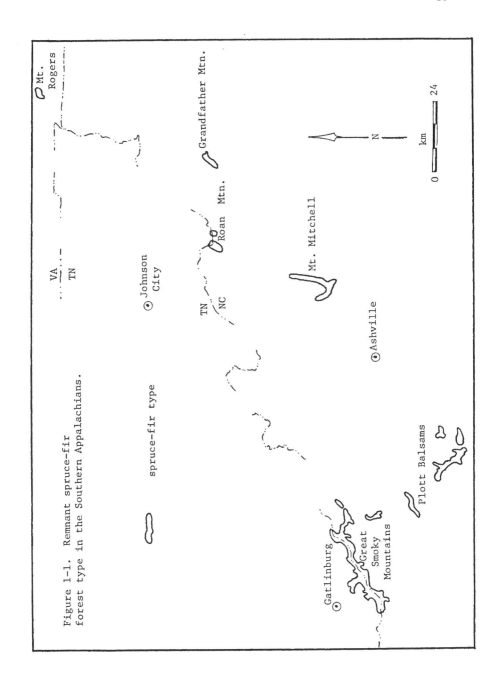

Figure 1-1. Remnant spruce-fir forest type in the Southern Appalachians.

spruce-fir type

the balsam woolly aphid in the Great Smoky Mountains National Park. From this loci, aphid dissemination was rapid throughout the northeast portion of the Park. The Great Smoky Mountains National Park was the last and largest reservoir of red spruce and Fraser fir in the Southern Appalachians (Reed, 1964). With a total of 14,164 hectares these forests comprised the majority of the remnant 20,234 hectares of the original spruce-fir forests. Enveloped in the protection of the National Park, the spruce-fir forests in the Great Smokies have been maintained in primeval condition, comprising a scenic and a scientific resource.

It has been 20 years since the initial infestation on Mt. Sterling. The aphid pest has colonized all the spruce-fir type, with some fir communities nearly 100 percent dead, others are alive but infested, and some remain uninfested. All possible elevations of fir, ranging from 1219 meters along the Tennessee side of the Transmountain Road to 2018 meters on Mt. Guyot have been exposed. All slopes, aspects and most other combinations of community and environmental factors possible within the Park have been exposed to balsam woolly aphids. This study was accomplished at a most expedient time in the development of the aphid within the Great Smoky Mountains National Park.

STUDY OBJECTIVES

The Great Smoky Mountains National Park contained the last Fraser fir that was highly visible to millions of visitors;

these are informed and concerned people that want to know what's happening in their environment. The obvious, drastic impact of the aphid on this scenic resource would attract significant attention.

Most previous work with the balsam woolly aphid and Fraser fir was in commercial forestry environments where silvicultural practices could modify the impact of the aphid. Species conversions, intensive site preparation, early harvest, and chemical sprays were possible. However, these are not applicable within the National Park Service. Rather the Fraser fir resource had to be evaluated in light of the balsam woolly aphid impact on the specific uses of that specific resource. Overwhelming questions were:

1) how will the existing resource respond to balsam woolly aphid attack?

2) will some Fraser fir trees survive?

3) will the balsam woolly aphid show preference for a particular type of environment or community? and

4) what might the future be for this resource?

Accordingly the following study objectives were developed by the University of Tennessee and the National Park Service.

13

1) DETERMINE LOCATIONS AND SEVERITY OF BALSAM WOOLLY APHID INFESTATIONS, REFERENCE MAY-JUNE, 1976.

Because the study area was so large and so isolated with Fraser fir being a disjunct component, it was necessary to define the resource and where it was threatened by the balsam woolly aphid. Aerial photography with seemingly endless ground checks and map interpretations accomplished this objective.

2) DETERMINE IF THE BALSAM WOOLLY APHID HAD SHOWN PREFERENCES FOR ANY OF THE MANY COMBINATIONS OF STAND SPECIES STRUCTURE OR COMMUNITY-ENVIRONMENTAL FACTORS.

After the resource was defined, sampling was designed to elaborate any trends in balsam woolly aphid preferences.

3) DETERMINE THE TIMING AND DEVELOPMENTAL SEQUENCE OF BALSAM WOOLLY APHID INFESTATIONS ON SELECTED PEAKS ALONG ELEVATION GRADIENTS.

This objective was implemented after initial sampling illuminated the dependency of balsam woolly aphid population development to stand elevation.

4) PROVIDE A SYSTEM OF PERMANENT PLOTS FOR FUTURE ANALYSES, REFERENCE THE 1976-77 STUDY.

Permanent plots were used in all cases to facilitate relocation and remeasurement in subsequent studies. Plots were marked in the field, on maps, and on aerial photographs.

14

5) RECOMMEND MANAGEMENT ALTERNATIVES COMPATIBLE WITH NATIONAL PARK SERVICE GUIDELINES THAT WILL MINIMIZE THE AFFECTS OF BALSAM WOOLLY APHID IMPACT ON THE RESOURCE.

It was never a study objective to determine balsam woolly aphid control techniques best suited to National Park Service lands. There have been numerous projects in the United States and Canada to release predators, to find parasites, to find effective chemicals, and to develop new chemicals or application techniques.

CHAPTER II

BIOLOGY OF THE BALSAM WOOLLY APHID

Of the two or three species of Adelges in North America, only
A. piceae has demonstrated a capacity to spread rapidly, maintain
epidemic populations, and cause fir mortality. Much of the avail-
able biological information on Adelginae has come from Canadian and
European research. The Southeastern Forest Experiment Station, U. S.
Forest Service formerly conducted research on aphid biology and con-
trol in the Southern Appalachians. Currently, the Forest Insect
and Disease Management Division of State and Private Forestry, U. S.
Forest Service remains active in aerial surveillance in the Southern
Appalachians.

Life Cycle and Morphological Development

The Adelginae subfamily is especially well noted for an
unusually complex life cycle, frequently including alternate host
plants. As an impressive example of the evolutionary flexibility
of these insects, Adelges piceae is thought to have evolved as an
asexual form of A. nusslini in the Caucasus with Picea orientalis
as the primary host. It lost its capacity for sexual reproduction
on alternate hosts when it moved to Europe. It now exists as an
exsule form capable of maturing and reproducing parthenogenetically,
thus maintaining its population entirely on the secondary fir host.

15

16

The functional life cycle of the balsam woolly aphid consists
of an egg stage, three larval instars and the adult (Balch, 1952):

1. Eggs laid by the stationary adult are attached
to the host bark by a silken thread. Initially
light amber-colored, the egg becomes orange-
brown as the embryo develops. Incubation is
about 12 days, but varies greatly with environmental conditions. Young larvae emerge head
first from the end opposite the thread attach-
ment, leaving the empty shell behind.

2. Motile larvae or "crawlers" are light purple with a
flattened, oval body between .35 millimeters and
.47 millimeters long. This is the first instar, or
"neosistens" stage. After inserting the stylet
(a slender, thread-like mouthpart used for feeding) into the host bark, the insect becomes dark
purple-black with a fringe of white waxy threads,
and the legs and antennae begin to atrophy. A
dormant period of variable length occurs before the
first moult. When this diapause occurs during the
winter, the neosistens is known as "hiemosistens",
while the corresponding phase of summer generations
is called an "aestivosistens". The hiemosistens
begins to feed when the sap rises in spring, while
the aestivosistens is dormant for three to eight
weeks. The insects remain stationary during the
following three larval moults.

3. The second instar develops after the first moult;
it has a longer (.45 millimeters to .55 millimeters)
and broader body than the first instar, with long,
curling waxy threads.

4. The third instar is similar to the second, except
that it is .60 millimeters to .86 millimeters long.

5. Adults are .70 millimeters to 1.0 millimeters long,
hemispherically shaped and covered with long, curling waxy threads of "wool". Since the insects tend
to congregate, the wool forms a dense protective
covering, making severe infestations readily visible.
All adults are female and reproduce by thelytokous
parthenogenesis, beginning with oviposition a few days
after maturation.

17

Seasonal Chronology

Metamorphosis of the balsam woolly aphid is influenced by temperature, moisture, and light. Temperature can be particularly critical; under controlled laboratory conditions, neosistens reared from the final adult generation of the season developed best with fluctuating temperatures. Progeny of the following spring developed with both constant and fluctuating temperatures and displayed a developmental variability that could provide a basis for distinct ecological races (Atkins, 1972). Such developmental variability allowed the balsam woolly aphid to infest hosts over a wide range of climatic zones and to adjust to local weather extremes; a portion of the population could break dormancy early if conditions were favorable, thereby increasing the number of generations per year. If early development occurred during unfavorable conditions, the slower developing portion of the population would maintain the infestation.

The number of generations per year is variable. Balch (1952) found two generations in Maritime Canada. In the Southern Appalachians the balsam woolly aphid overwinters as a dormant neosistens and the hiemosistens break dormancy in mid-March, as elsewhere. However, development here is generally faster throughout the season and it continues into early fall. Amman (1962) found that three and sometimes four generations could reach maturity during a season.

Intraregional variations in seasonal biology are greatest in the Pacific Northwest, due to the wide range of elevation and different host species. At the higher elevations, where the balsam woolly aphid is found on Pacific silver fir and subalpine fir, only two generations were observed. At intermediate elevations there were three generations

annually, and woolly aphids on grand fir in the valleys had as many

as four generations (Mitchell, Johnson and Rudinsky, 1961).

Variations in number of generations per year also occur within

localized areas as a result of stand conditions. Insects on open-

grown or stand-edge trees developed faster, perhaps by a full month,

than those on trees in the stand interior. In dense stands those

insects located near the base of the tree may experience fewer

generations than those higher on the stem because of slower develop-

ment (Mitchell et al, 1961).

Dispersal

Passive dissemination of insects is primarily determined by

abundant availability of a mobile stage, plus the direction and

velocity of wind. Dispersal of the balsam woolly aphid occurs during

the egg and crawler stages and is mainly passive. Once dislodged

from the tree, they can be carried more than 300 feet by surface winds

and several miles by vertical air currents (Balch, 1952). The crawlers

can move about on the bark in search of a suitable attack site, and

they also jump. Gravity becomes important in infesting advance repro-

duction.

Synchronous dispersal and arrival of many aphid crawlers on the

host may facilitate establishment and survival. In laboratory

experiments, groups of insects settled more rapidly than isolated

individuals (Atkins, 1972). In the forest, adjacent heavily infested

and uninfested trees are common. Since the matted wool masses are

protective, groups may readily become established and survive, at

least up to the point where overcrowding occurs. Infestation spread

·

would be affected by the number of generations per year and perhaps
also by the synchrony of crawler development and dispersal, both of
which are influenced by environmental factors (Atkins, 1972).

Population Dynamics

Factors that limit the distribution and abundance of the balsam
woolly aphid are varied, but they can be organized as biotic, abiotic,
and applied.

Abiotic factors. Mortality due to climatic factors may be high
occasionally, but they do not provide an effective limit to population growth (Balch, 1952). Of the various climatic factors, winter
temperatures appear to be particularly important: the lethal minimum
temperature for overwintering neosistens is around -34^0 C and -18^0 C
for all other stages. Low temperatures during the summer months increase incubation time. In the Southern Appalachians winter weather
is not severe enough to effectively limit balsam woolly aphid populations. Even -33.8^0 C for one night had no affect on resident aphids
on Mt. Mitchell (Amman, 1967).

High temperatures appear to have little effect on aphid survival
except in combination with low relative humidity; dessication is a
frequent cause of mortality on exposed bark areas, especially during
late summer. Heavy rains may mat the wool and lead to subsequent
dessication, or wash eggs away, but neither rain nor heavy fog will
"drown" aphids. Protection from the weather is afforded by the wool
itself, the shade of the tree crown, and by the insects habit of

congregating under twigs, branches, mosses and lichens and in bark
crevices and lenticels.

Balch (1952) observed that climatic factors were important
influences on the rate at which the population reached outbreak
proportions and dispersed. Winter temperatures can produce irregular., periodic fluctuations in the population over large areas. The
moderate weather conditions in the Southern Appalachians facilitate instead of control balsam woolly aphid development.

Applied factors. At present, applied control is impractical in
the forest. Lindane is effective in killing the balsam woolly aphid,
but it must be applied until run-off occurs along the entire bole
and branches. To further compound the difficulty of this procedure,
spraying must be done from the ground. Silvicultural control has
involved complete removal of the host species.

Biotic factors. Since reproduction is not limited by the
necessity of mating, the biotic potential of the balsam woolly aphid
is quite high. The average number of eggs laid by one adult of the
hiemosistens generation is 100, although as many as 250 have been
observed. The less fecund aestivosistens generations average 50 eggs.
Thus the potential annual multiplication approximates a magnitude
of 5000, and under optimal environmental conditions, survival to
the adult stage is about 60 percent (Balch, 1952).

A greater proportion of the winter generation survive environmental stresses than the summer generation. Amman (1970) observed

that during the initial period of outbreak in a stand, increased insect population density appeared to result in increased fertility. Later there was a strong inverse relation between adult density and fertility, probably due to competition for food at peak population densities. Survival and reproduction were best on trees with less than extremely dense populations of aphids.

The balsam woolly aphid is apparently not bothered by parasites or disease, and the reproduction potential is so high that predators cannot significantly reduce the populations before the host trees die. Amman (1970) recorded ten native and one introduced predators in North Carolina, most of which were mites feeding on the egg stage. Predators introduced to control aphid populations have either failed to become established or have otherwise been ineffective.

In the absence of other biotic factors capable of effective control, the upper limit of the population must be determined by intraspecific competition and starvation. In the case of the balsam woolly aphid, this is accomplished by severe injury to the host, usually followed by its death. Thus the host tree becomes the most significant factor in determining aphid population levels. Balch (1952) recorded the following population density sequence on stands of balsam fir in Maritime Canada.

1. Insects become established on individual scattered trees (often the larger trees with rough bark) and spread to neighboring trees.

2. Population increases steadily (unless depleted by severe winter temperatures) and reaches a peak

when a large proportion of trees within the stand
are heavily infested.

3. Populations decline as susceptible trees are
    eliminated, and remain at lower but continually
    fluctuating levels with recovery, reinfestation
    and occasional new infestations occurring simul-
    taneously.

Characteristics of fir stands and individual trees that were
most relevant in the development of an infestation were a major
portion of this project.

Activity

Balsam woolly aphids in the crawler stage are active and usually
move about for several hours before inserting their sylets. If
suitable feeding sites are available most will settle near their
parent, but they are capable of moving 30 meters under their own
power; they can survive for at least 8 days without feeding (Balch,
1952). However, crawler mobility is more important for movement on
the host tree than for dispersal. Distribution on a given host de-
pends upon available light and gravity, plus accessibility of young
parenchyma at wounds, lenticels and crevices, or feeding areas stimulated by other larvae.

The initial locus of stem infestation by wind-transported aphids
is determined by chance primarily, although Amman (1970) observed
that it usually was near mid-height. This varies with tree height

and age; on older trees with rough, thick bark, infestations start
higher in the tree, but on young trees infestations usually begin
at the base. Woolly aphids migrate both upward and downward after
initial attack. The selection of a certain height on the bole is
probably a result of the crawlers tendency to settle in moderately
intense light (Balch, 1952), but bark thickness and bark texture
also seem to be important. Crawlers are mobile and selective
enough to locate preferred feeding sites.

The feeding process begins when the stylet is inserted intercellularly (aided by a dissolving action of the saliva on the middle
lamella), passing through the epidermis into the cortical parenchyma.
During feeding, the stylet is partially withdrawn and reinserted
several times in new directions, but the insect itself remains stationary. Saliva flows into adjoining tissues in the bark, spreading
to the cambium and the xylem. Nerves at the base of the stylet provide a tactile sense; obstructions can be felt, absence of satisfac-
tory food detected, and relocation of stylets prompted. Spring feeding usually begins first on more vigorous trees with earlier sap
flow; inception of feeding may vary among individual insects on a
given tree by as much as 20 days (Balch, 1952).

HOST/PEST INTERACTIONS

The balsam woolly aphid is an obligate parasite on Fraser fir,
obtaining food and shelter in amounts dependent upon the hosts capacity to provide them. There is, considerable evidence that some trees

are more favorable to the multiplication of the insect than others.
The ability of the woolly aphid to become established and reproduce
is affected by external conditions of the tree, e. g. bark characteristics, form, size, and by feeding conditions within the bark.
Such qualities can be influenced by genetics, modified by the environment, and/or related to the age, size, and physiological condition
of the tree (Balch, 1952).

Microscopic Changes

Microscopic tissue changes are similar in all infested Abies;
the differences exist mainly in the rate and the degree of change.
Damage can develop from physical injury caused by insertion of the
stylet but the chemically-induced injury produced by the salivary
secretions is by far the most important because the secretions
change the balance of growth hormones and inhibitors in the host,
causing abnormal development of wood and bark tissues (Balch et al,
1964). Individual feeding actions of such minute insects would
hardly interfere with normal host functioning, but their enormous
capacity to reproduce and rapidly spread disastrously multiplies the
effects of each balsam woolly aphid.

Balch (1952) made the first comprehensive histological studies
of the balsam woolly aphids feeding process and its effects on host
tissue. He observed that the stylets were inserted intercellularly
and seldom, if ever, penetrated the cell wall. They pass through
the epidermis of phellem into the cortex or phelloderm and feeding

takes place only in the parenchyma. In young shoots, however, the
sieve cells are sometimes penetrated slightly. The insertion of
stylets is accompanied by injection of a salivary substance which
occasionally flows into adjoining intercellular spaces.

A substance contained in the saliva, or produced in the cortical
tissue in reaction to the saliva, diffuses from the point of insertion and causes parenchyma cells contiguous to the stylet tracks to
increase in number and size. Concurrently, enlargement of the nuclei
and thickening of cell walls occurs; giant cells, six or seven times
larger than normal are produced. Such parenchyma volume disrupts
phloem sieve cells, resulting in an impediment to carbohydrate
movement.

By the following season, neighboring cells proliferate and
form a secondary phellogen or periderm surrounding the pockets of
abnormal tissue. This represents the initial stage in the wound-
healing process. European firs are able to complete this process
rapidly enough to isolate affected tissues; some North American firs,
notably white fir and noble fir exhibit this resistance while other
natives apparently do not. Tree vigor may cause some differences
in individual resistance (Bryant, 1974).

The salivary secretion also affects the xylem; cell division
is stimulated in the cambium, resulting in production of tracheids
with thickened, irregularly-shaped cell walls and dark, hard, brittle cellulose. These tracheids resemble those of compression wood:
secondary walls are marked with checks, the number of rays and

parenchyma cells increase, cells are highly lignified, short, and thick-walled with a small lumen. This abnormal wood is called "rotholz" because of its reddish color (Doerkson and Mitchell, 1965).

The reaction of the cambium is dependent upon the vigor of the tree and the intensity of infestation; the greatest amount of abnormal wood is produced in fast-growing stems that are moderately infested (Balch, 1952). Presence of rotholz can be useful in dating infestations, but its absence does not necessarily indicate that an attack has not occurred. It is only found in certain areas of affected wood; even a heavy attack will not always produce rotholz and it is not uniform around a growth ring or along the stem vertically.

Aphid damage apparently modifies the bark so that a greater proportion of insects can survive per unit bark area. The bark of trees with increasing levels of infestation has higher amounts of protein, while those with declining aphid populations have reduced protein content. Growth rate also influences bark formation and thus may be relevant to aphid survival. Amman (1970) suggested that the bark of vigorous trees should logically provide the most food for aphids and hence support the largest populations. He proposed that the aphid population which trees could ultimately support could be predicted on the basis of growth rate.

Death of the host is caused by a combination of factors:

1. production of abnormal cells and resultant interference with translocation of fluids

through the xylem,

2. killing of the outer tissues of the bark
    by the toxic effect of the saliva in
    heavy infestations, and

3. production of secondary periderm in
    sufficient quantities to interfere with
    respiration in underlying tissues (Balch,
    1952).

Macroscopic Changes

Effects of aphid attack that are outwardly apparent are more
variable among host species than microscopic tissue changes.
Infestations are usually classed as stem attack or crown attack,
depending on where the aphid population is concentrated. Greenbank
(1970) suggested that the type of attack is influenced more by climate than by insect or host species characteristics; stem attack was
more common in continental climatic zones and crown attack was more
common in maritime zones. Either or both may prevail in transitional
zone forests. The bracted balsam fir of Virginia usually suffers
crown attack; stem attack is the most frequent on Fraser fir, although
small suppressed trees and reproduction can develop crown gouting as
aphids drop onto them from the overstory (Amman and Talerico, 1967).

Crown infestations produce a swelling at the nodes (Figure 2-1),
then a deformation of shoots and internodes followed by down-curling
branches, tip inhibition and die-back. Appreciable mortality does
not occur until after 10 to 20 years, and growth recovery is possible

Figure 2-1. Gouting of the nodes caused by balsam woolly aphids on a Fraser fir seedling. Much of the advance reproduction growing under an infested canopy showed gouting. Occasionally, these seedlings will resume normal growth patterns after the overstory infestation ceases.

if the aphids are controlled (Bryant, 1970).

Stem infestations cause death quite rapidly, often within a few years of initial attack, because of the direct translocation impediment. Stem attack is evidenced by wool spots on the stem and changes in foliage color from the normal healthy blue-green to a faded yellow-green, then bright rusty-red and finally dead-brown. The speed of these changes seems to reflect the severity of attack and/or the lack of tree resistance. Amman (1970) observed the following infestation pattern, regardless of host characteristics:

1. increased diameter growth at the time of initial attack,

2. increased aphid survival as the bark is changed by aphid feeding,

3. peak aphid population, with aphid survival closely related to tree condition,

4. decline in aphid survival and also tree growth,

5. disappearance of aphids, and

6. death of host.

Other detrimental effects of infestation include increased susceptability to _Armillaria_ root rot and impairment of reproductive functions. Hudak and Wells (1974) found that damage caused by woolly aphids was primarily responsible for the high incidence of _Armillaria_ root rot in aphid-damaged stands, regardless of site quality.

31

Fedde (1973) found that cone production decreased rapidly and was
restricted to the upper third of the crown in aphid-infested Fraser
fir trees. In addition, cones were shorter by 25 percent, more
brittle, and less impregnated by gum-like substances. Seeds were
smaller than normal, frequently lacked megagametophyte, and were
likely to be infested by seed chalcids, thereby maximum importance
for fir regeneration must be assumed by the advanced reproduction
in the understory. Clearly new seedlings will be scarce in an
infested stand.

CHAPTER III

PROJECT RESEARCH PROCEDURES

Color infrared aerial transparencies covering areas of potential
balsam woolly aphid infestation were provided by the U. S. Forest
Service in cooperation with the National Park Service in May, 1976.
Analysis and procedural organization included locating transparency
frame center on maps, identifying the spruce-fir type, and catego-
rizing aphid infestations.

Frame Centers

Frame centers were located on 1:24000 topographic maps. Align-
ment of the transparency on the map served as a reference when
transferring features from the transparency to various maps. The
determined flight lines and the extent of aerial coverage showed
the areas that needed thorough ground reconnaissance. This information will provide valuable future uses for these aerial transparencies.

Forest Types

Spruce-fir types were identified and mapped as stand units on
1:24000 topographic maps. The apparent spruce-fir forest type lines
were transferred to the topographic maps to aid in field interpretation and plot lay-out. The obvious contours on the transparencies
facilitated separation of stand units and their transfer to the

maps. Stand units were based somewhat upon topography, aspect, size, and accessibility. It was not our intention to provide a cover-type map based exclusively upon community characteristics of species composition and age structure.

This technique was invaluable and prerequisite to designing sampling schemes and to identifying balsam woolly aphid infestation hotspots.

Infestation Centers

Balsam woolly aphid infestations were located and classified into infestation categories according to apparent insect activity. Each stand unit was identified and numbered according to the following scheme:

1. <u>Active</u>: fir dead and dying from apparent balsam woolly aphid attack.

    a. Light: small infested area or scattered affected trees.

    b. Medium: well-developed infestation with some dead trees in the center.

    c. Heavy: severe, well-developed infestation with many dead trees in a large total area.

2. <u>Dead</u>: high fir mortality with no "hot" trees (trees just beginning to decline showed bright yellow-green, while dead trees were gray).

    a. Light: scattered small patches of dead trees, perhaps not caused by balsam woolly aphid attack.

b. Medium: larger patches of dead

trees.

c. Heavy: complete mortality over

large areas.

3. <u>None</u>: no evidence of balsam woolly aphid infes-

tation.

The entire spruce-fir type was divided into 11 large geographic

areas which permitted grouping of areas that had experienced similar

levels of aphid activity in terms of length of time, size, and degree

of damage. Geographic areas were numbered in order beginning in the

eastern portion of the spruce-fir distribution. Stand units within

the geographic areas were numbered and these codes were used as

identification in data analysis.

Stand unit sample selection was made randomly from each infesta-

tion category in proportion to the total possible. Seventeen stand

units were sampled with 102 plots, selected from the 11 geographic

areas, during the 1976 field season. Sampling accomplished during

the second field season varied slightly and it will be described later.

FIELD WORK - 1976

<u>Plot Location</u>

Community and individual tree characteristics that were thought

to be related to balsam woolly aphid infestations of Fraser fir were

evaluated in 1976. Six plots were established in each stand unit.

Potential plot sites were located in advance on the topographic maps

to assure representative ranges of slope, aspect, and infestation

intensity. Sometimes selected plot sites were modified in the field to avoid excessively dense rhododendron thickets or steep terrain. Actual plot locations were recorded on topographic maps and aerial transparencies.

After plot center was established, ten meters square were delineated using measured north-south and east-west diagonals. Four trees of promising longevity were marked with numbered aluminum tags. positioned about one meter above ground facing plot center. Bearing and distances from plot center to each witness tree were recorded. Black and white photographs that could be used to describe each plot as well as aid plot relocation were made.

Each plot and the general area surrounding that plot were described subjectively to typify the Fraser fir community that existed in 1976. Information collected included the overall level of balsam woolly aphid activity based on subjective analysis, general stand structure and species composition (both past and probable future relationships), the dominant ground cover (tree seedlings, Vaccinium, Rhododendron, Viburnum, ferns, Oxalis), the gouted condition of fir seedlings located under an infested overstory, and any other pertinent characteristics for future analysis.

The type and extent of past disturbances, predominantly man-oriented, were recorded. Types included logging, fire, multiple-tree blow down, plus other means of initiating secondary succession within the community. Plots were classified as either disturbed or not disturbed.

36

Objective measurements made on each plot included the following:

1. Species composition and diameter frequency: Balch
   (1952), Amman (1970), and Greenback (1970) thought
   diameter was at least indirectly relevant; Greenback (1970) correlated bark characteristics to
   diameter. Frequency of all trees greater than
   1 centimeter diameter at 1.37 meters was recorded
   by diameter class. A metric diameter tape was
   used for all measurements.

2. Slope: percent slope was measured with an Abney
   level.

3. Aspect: compass bearings at right angles to the
   slope were made from plot center.

4. Elevation: contour lines on a 1:24000 topographic
   map were used to estimate plot elevation.

5. Plot BWA infestation rating: all fir trees greater
   than 4 centimeters diameter were numbered with
   temporary tags. Each tree was classified as to aphid
   infested, uninfested, plus healthy, dying, or dead.
   Aphid presence was determined at 1.37 meters along
   the stem. Health evaluations were based upon crown
   color.

Subsample of Individual Tree Characteristics

Based upon the temporary numbered tags, a random sample of three

37

Fraser firs was selected from each plot for detailed analysis of
the relationship between balsam woolly aphid infestation and individual tree characteristics. Specifically, individual tree
characteristics that might be related to aphid attack and indications of aphid attack intensity were sought.

Height. Balch (1952) suggested tree height as a possible
factor in the susceptibility of balsam fir to balsam woolly aphid
attack, perhaps because taller trees were more likely to intercept
the air-borne eggs and crawlers. Total height was measured with
an Abney level.

Crown class. Schooley and Oldford (1974), Balch (1952), and
Johnson et al (1963) correlated crown class with aphid attack.
Crown classes were defined according to the Society of American
Foresters Forestry Terminology as suppressed, intermediate, codomi-
nant, and dominant.

Bark texture. Irregularities in bark texture provide shelter
and feeding sites for the balsam woolly aphid (Balch, 1952), and
may also reflect tree age, vigor, and stand density. Each tree
was rated for bark texture after inspection:

1. very smooth,

2. some cracks, crevices and other irregularities,

3. more cracks, crevices and other irregularities, and

4. very convoluted, many irregularities.

Bark epiphytes. Moss and lichen cover are more abundant in stands
that are stagnated or on very moist sites (Crandall, 1957). Bark

epiphytes may protect trees by prohibiting woolly aphid feeding; conversely, aphids might use the epiphytes for protection. Previous
studies have not examined this relationship.

Each tree was placed into one of four bark epiphyte categories:

1. clean bole, no epiphytes,

2. sparse epiphyte cover,

3. light but uniform or fairly heavy in patches, and

4. uniformly thick moss and lichen cover.

Bark thickness. Since the woolly aphid stylet must penetrate
the outer bark, bark thickness could be related to aphid attack.
Bark thickness was measured directly from the increment core with
a millimeter scale.

Increment cores. Increment cores were needed to determine age
and growth rate. The cores were taken at breast height along north-
south axes; when possible the entire diameter was sampled, but for
larger trees two cores were obtained at 180 degrees. This procedure
provided two pith-to-bark samples of each tree, which were averaged
to calculate growth rate.

In preparation for analysis, increment cores were glued between
grooved slats of redwood and cut with parallel saw blades to produced
a stable, mounted cross-section. Cores were examined with an 80 X
binocular microscope. Total age was obtained by counting annual rings
from cambium to pith. Annual increment was recorded to the nearest
0.25 millimeters for each year from 1976 to 1967, and also for the

years 1962 and 1957.

Aphid population on bole. Indications of aphid attack intensity were necessarily subjective. Estimates of aphid population were made with an emphasis on consistency and guided by the Amman (1969) method; he used an 8 X 8 centimeters plastic grid placed on the tree trunk at breast height to supplement an overall subjective appraisal. The grid method alone was not sufficient because woolly aphids sometimes congregate on one side or at various locations along the vertical axis of the bole.

On the basis of overall inspection and use of the grid, each tree in the subsample was placed into one of four aphid population categories:

1. none,

2. light - less than four wool masses per grid square,

3. moderate - four masses per grid square to 25 percent covered, or

4. severe - more than 25 percent covered with wool masses.

In 1977, a fifth aphid population category was added to reflect those living trees that no longer supported aphids. Such trees gradually deteriorated over extended periods after the aphids were gone. Aphid population ratings from the two field seasons were standardized before analyses.

Crown color. Foliage color was used as a symptomatic criterion for determining aphid infestation severity. Each subsample

tree was rated according to the following crown color categories:

1. green - no discoloration of leaves,

2. light-fading - yellowing or reddening of some leaves
   or branches,

3. red-dying - bright red leaves throughout the crown, and

4. brown-dead - dark red to brown leaves throughout the
   crown.

FIELD WORK - 1977

Observations made during the 1976 field season indicated that
balsam woolly aphid infestations occurred first at the northern
hardwood-fir ecotone and spread toward the summit. To investigate
this relationship, changes in sample selection and plot location
were implemented during the 1977 field season; plots were located
along elevation transects at right angles to the contours.

Sample Selection

Sample selection was restricted to selected mountains that
offered the greatest elevation change within the distribution of
Fraser fir. In addition to insuring an adequate range of species
composition and stand age structures, mountains were selected
throughout the spruce-fir zone in the Great Smoky Mountains National
Park to obtain a representative sample of length of infestation
time. Mountains sampled were Big Cataloochee, Mount Guyot, Mount

41

LeConte, Mount Mingus, and Clingmans Dome.

Big Cataloochee has been exposed to balsam woolly aphid infestations for the maximum time, while Clingmans Dome has only recently been infested at the lower elevations of fir. Therefore, this selection of mountains provided for the study of aphid infestations over their entire developmental sequence.

Transect and approximate plot locations were drawn on the topographic maps prior to field sampling. Each mountain was sampled as representatively as possible: aspect, slope steepness, stand species structure, and spur ridge location were all considered.

Plots were located at 60 meters elevation intervals along the transects, being measured on topographic maps and in the field. A pocket altimeter was used to substantiate the elevation of each plot.

Plot evaluation was accomplished as in 1976, except the fir subsample was deleted.

CHAPTER IV

DISTRIBUTION OF THE APHID IN THE
GREAT SMOKIES AND THEIR
IMPACT ON FIR TREES AND COMMUNITIES

This chapter is a composite of results from related but
distinct studies that span the entire project. Information gathered in the early studies influenced the collection format for information from subsequent work. Our knowledge and understanding
of the balsam woolly aphid and its impact upon Fraser fir also
increased as the project developed, therefore, much of what was
ultimately accomplished reflects changes made in the initial de-
signs.

Data modifications were made before analysis was accomplished.
For example, since the aphid population rating on individual trees
was derived somewhat differently in 1976 than in 1977, they were
standardized before analysis. Data were also transformed prior to
analysis. Individual trees were the field observations, but the
analysis required that individual plots be treated as observations.
Therefore, the sum of individual tree diameters within each aphid
population category was used in analysis; diameter was directly
related to those factors that facilitated aphid feeding.

42

43

DATA ANALYSES

Community - Environmental Factors Analyses

Canonical correlation analyses were used to determine which community and environmental factors were associated with the various aphid population levels and stages of infestation development. Canonical correlation is a form of multivariate analysis that enables determination of the degree of relationship between two groups of variables. This is accomplished through two linear combinations of coefficients (weights), such that the Pearson product-moment correlation between the two groups of variables is as large as possible. Additional combinations of coefficients can be calculated as long as the new combination is uncorrelated with previous combinations. There can be as many linear combinations as the minimum variables in either group. However, each subsequent linear combination accounts for less variation between the two groups of variables. These linear combinations are also called "canonical variables."

Two primary statistics were calculated through canonical correlation analysis. The first statistic consisted of a between group correlation coefficient (canonical-R) showing the overall strength of the relationships found for that grouping of linear combinations (canonical variables). The canonical-R for each new linear combination of the variables is lower than the preceding one. These correlation coefficients were tested for significance with Chi-square

at the probability of one percent. The second statistic was the within group correlation coefficient which expressed the degree of relationship between the variables within the two groups. These coefficients also represented how the variables in the first group were related to the variables in the second group.

For this study one group of variables was composed of the aphid population levels expressed as the sum of diameter within each category and the other group consisted of the community and environmental factors. The aphid population levels were compared with the community and the environmental factors combined and with them independent of each other. Separate analyses were performed to reduce the random variability when all factors were grouped together, thereby identifying realistic relationships between the aphid and its host.

Separate analyses were accomplished with the following data:

1. the 1976 data--84 plots total,
2. the 1977 data--47 plots total,
3. the 1976 and 1977 combined data--131 plots total, and
4. the combined 1976 and 1977 data from the western geographic areas--81 plots total.

The western portion of the spruce-fir type, i.e., the Clingmans Dome area, the Mt. LeConte area, and the Mt. Kephart area, represented the most recently infested portion of the Park. Separate analyses of these areas presented a determination of the factors

associated with the earliest stages of balsam woolly aphid infestation activity. Therefore, a total of 12 canonical correlation analyses were accomplished; three groupings of the community-
environmental factors within each of four groupings by year and
location.

Canonical correlation analyses were not solely sufficient to
identify the relationships between aphid infestation levels and
the community and environmental factors. For example, was the correlation between a particular aphid infestation level and the percentage of fir in that community real or was it due to the correlation of the community structure at that elevation and the aphid
population level? These interrelationships were tested using standard correlation analyses to produce a correlation matrix of the
community and environmental factors.

Data were summarized by species and aphid population level
categories on a WANG calculator. Canonical correlation analyses and
standard correlations were performed on an IBM 360/65 computer using
Statistical Analysis Systems programs (Barr et al, 1972; Barr et al,
1976).

Individual Tree Analyses

Relationships of individual trees to balsam woolly aphid infestation were analyzed with Statistical Analysis Systems (Barr et al,
1976). Continuous variable data, e.g., bark thickness, diameter,
height, age, growth rate, slope, aspect, and elevation, were sorted

according to categories of aphid population levels and crown color; means, frequencies, and variances of each were calculated. The Scheffe multiple range test (Chew, 1976) was used to compare each variable by aphid population and crown color categories.

Ranked data included bark epiphytes, bark texture, crown color, aphid population, and crown class. The following characteristic combinations and comparisons were analyzed with Chi-square contingency tests.

1. crown color with aphid population

2. crown class with aphid population

3. crown class with crown color

4. bark epiphytes with aphid population

5. bark epiphytes with crown color

6. bark texture with aphid population

7. bark texture with crown color

The same analyses were accomplished with the combinations grouped by geographic units to provide comparisons along time of infestation gradients.

DISTRIBUTION OF THE BALSAM
WOOLLY APHID IN THE GREAT SMOKIES

The balsam woolly aphid is distributed throughout the spruce-fir vegetation type in the Great Smoky Mountains National Park. Figure 4-1 in the attached map pocket presents the aphid distribution, reference July, 1977. The earliest infestations were on

47

Mt. Sterling and from that locus the insect has spread throughout the available host range. Fraser fir mortality was most severe in the eastern portion of the Park with varying amounts of mortality occurring throughout the remaining host distribution. That the aphid was intermediate in its chronological development and range expansion within the Great Smoky Mountains, provided an excellent opportunity to study those factors associated with aphid infestations in Fraser fir.

Mt. Sterling Geographic Area

The Mt. Sterling Geographic Area extended from Mt. Sterling southwest along Mt. Sterling Ridge to Big Butt. Nearly 100 percent of the fir within the canopy had been killed by the aphid. See Figure 4-2. Along Mt. Sterling Ridge near the junction with Mt. Sterling Gap Trail, there were two living, infested firs that probably owed their continued presence to their open-grown habit; full, broad crowns extending nearly to the ground provided added vigor which extended their lives. A dense, even-aged patch of small, pole-sized fir located on top of Big Butt showed no mortality, however, ground checks confirmed aphid presence on some trees.

Big Cataloochee Geographic Area

Being contiguous with the Mt. Sterling Area on the east, the Big Cataloochee Geographic Area extended along the Balsam Mountains

48

Figure 4-2. View from the summit of Mt. Sterling southwest toward Big Cataloochee. The living conifers are all red spruce; all the fir have been killed by the balsam woolly aphid.

northwest to Tricorner Knob and south to near Beech Gap. Fir mortality was near maximum along the lower slopes of the Balsam Mountains, however some individuals remained alive and infested along the ridge-line. Balsam woolly aphid development on Mt. Hardison was apparent from the summit of Mt. Guyot.

Heintooga Geographic Area

The Heintooga Geographic Area was discontiguous with the Cataloochee Area, extending further south along the Balsam Mountains from Spruce Mountain to Polls Gap. Fir mortality was extremely heavy in the Spruce Mountain portion of the area, and balsam woolly aphids have infested most of the fir stands at the lower elevations throughout the entire area.

Mt. Guyot Geographic Area

The Mt. Guyot Area was bounded on the northeast by Camel Hump Knob, and on the southeast by Tricorner Knob; Greenbrier Pinnacle and Guyot Spur on the Tennessee side of the State-Line Ridge were included. Fir mortality was observed throughout the lower elevations with the heaviest damage on the Tennessee side of Old Black, Inadu Knob and Snake Den Mountain. Heavy mortality was observed in the Ramsey Prong Drainage during the 1977 field season, although very little mortality or indication of aphid presence was observed there either on the aerial photographs or during ground checks in 1976. Mt. Guyot at 2018 meters elevation is the most prominent peak in the area, and it was ringed with balsam woolly aphid

infestations on all sides to about 1951 meters elevation. Aphids
were observed on a single fir located at the very summit of Mt.
Guyot, however careful checks surrounding it failed to locate additional infested trees.

One plot at 1951 meters on the west slope of Mt. Guyot contained two suppressed trees that had an isolated balsam woolly aphid on
each stem. These trees were growing subordinate to a large dominant which had probably intercepted aphids moving on upslope winds.

Dashoga-Hyatt Ridge Geographic Area

This Area extended from Mt. Hardison near Tricorner Knob south
along Dashoga and Hyatt Ridges, the two prominent ridges in the
area, to approximately Hyatt Bald at 1571 meters elevation. With
high aphid populations, fir mortality was most severe in the southern portion of this Area, i.e., around Breakneck Ridge. Mortality
was most prominent at the lower elevations throughout the entire
area.

Mount Sequoyah Geographic Area

The Mount Sequoyah Area extended along the State-Line Ridge
from Tricorner Knob to Eagle Rocks. The size of the infestations
in this Area were smaller than in previous areas and fir mortality
was less pronounced. The largest infestations were located south
of the State-Line Ridge between Mt. Sequoyah and Mt. Chapman.
The aphid was distributed throughout the lower portion of the fir

range in this location, but had not yet reached the top of the

Ridge. Mortality was restricted to the lower elevations of the

State-Line Ridge, balsam woolly aphid infestations were widely

scattered along the hardwood-fir ecotone.

Hughes Ridge Geographic Area

The Hughes Ridge Geographic Area was entirely within North

Carolina, extending from Pecks Corners on the State-Line Ridge

south encompassing the entire range of Fraser fir in that locali-

ty. Due to the prevalent lower elevations (generally less than

1676 meters), there were few pure stands of Fraser fir. Inter-

pretations of aerial transparencies were more difficult. How-

ever, ground checks revealed that most of the fir located along

the hardwood-fir ecotone had already been killed by the aphid.

There were many small infestations scattered throughout, with

one large infestation on Katalsta and Bulldie Ridges.

Laurel Top Geographic Area

The Laurel Top Geographic Area extended southwest along the

State-Line Ridge from Pecks Corners to Charlies Bunion. Pure fir

stands were infrequent in this Area, being found only on the summit

of Laurel Top and Porters Mountain. The southern portion of this

Area had only two small infestations, located about 800 meters

east of Laurel Top and just below the State-Line Ridge. The

northern portion of the Area had many small, scattered infestations

just above the hardwood-fir ecotone that contained dead fir trees. A spur ridge north of Laurel Top, known as Woolly Tops Mountain, contained the heaviest aphid infestations in this Area.

Mount Kephart Geographic Area

This Area was bounded by Charlies Bunion on the northeast and Newfound Gap on the southwest, crossing the Boulevard between Anakeesta Knob and the Jumpoff. Heavy infestations of moderate size were located south of Masa Knob and along the ridge south and east of Icewater Springs. There was some fir mortality in these infestations and they were confined to the lower edges (less than 1646 meters) of the spruce-fir type. Numerous and smaller infestations were distributed on both the north and south side of the State-Line Ridge between Mt. Ambler and Newfound Gap. These infestations were mostly restricted to the hardwood-fir ecotone. The aphid was observed along the State-Line Ridge during the 1977 field season; some trees at Newfound Gap were heavily infested.

Mount LeConte Geographic Area

Because of its unique geographic placement and stage of aphid infestation development, Mt. LeConte provided excellent study opportunities; the entire mountain is located north of and connected to the State-Line Ridge by the Boulevard which provided a continuous band of Fraser fir between Mt. Kephart and Mt. LeConte.

Mt. LeConte was completely encircled by isolated infestations of

varying sizes. The largest infestations were located on northern aspects. These infestations were no longer restricted to the lower elevations where most of the fir had already been killed, but they were found as high as 1829 meters. Infestations south and east of the summit were small and widely scattered, with only a few dead fir.

Several of the large, planted Fraser fir landscaping the Mt. LeConte Lodge were found to be infested in July, 1977. Careful ground checks at that time did not reveal additional infestations, however, by November more aphids were on the previously infested trees plus numerous other trees surrounding the Lodge were found to support isolated but frequent aphids. This area was situated at the head of a long, deep valley which serves to funnel air currents toward the summit, thereby providing a ready invasion mechanism for balsam woolly aphids to the summit of Mt. LeConte.

Clingmans Dome Geographic Area

At the time of aerial photography, this Area showed little indication of balsam woolly aphid activity. Consequently it was designated from Newfound Gap to Silers Bald including the various spur ridges from Mt. Collins and Clingmans Dome.

Aerial transparencies showed several small scattered infestations along some of the spur ridges; e.g., Forney Ridge off Clingmans Dome and Loggy Ridge from Mt. Buckley. Extensive ground checks revealed numerous and large infestations in these areas. In most

cases, known infestations in the Clingmans Dome Area were at the hardwood-fir ecotone, however some infestations have spread almost to the State-Line Ridge along Mingus Lead and along Big Slick Ridge. A large infestation with severe fir mortality was located on Noland Divide. This infestation was still below 1768 meters, reference 1977 field season. Balsam woolly aphids were not found on the summit of the Dome during November, 1977.

APHID IMPACT ON FRASER FIR

Canonical correlation analyses results are presented in Tables 4-1 through 4-4. Due to the analysis techniques utilized in these experiments the results will be presented and discussed in terms of "typical" communities. These typical communities represent a composite, as determined by analysis, according to the various levels of balsam woolly aphid population levels and community-environmental factors. Cognizance must be taken that a specific community is not being described, rather a typical community as defined by analyses.

Aphid population levels (APOP) as coded in the field were used in analysis to define "typical stands" as follows.

1. The typical uninfested stand was based on a very strong positive APOP-1 correlation with the other APOP levels very weakly and/or negatively correlated.

56

TABLE 4-1

Canonical Correlation Coefficients for Balsam Woolly
Aphid Populations (1976 plus 1977) Throughout the
Spruce-fir Type in the Great Smoky Mountains (131 plots).

	ALL FACTORS		ENVIRONMENTAL FACTORS ONLY		COMMUNITY FACTORS ONLY
CANONICAL VARIABLE	#1	#2	#1	#2	#1
CANONICAL R	.970	.581	.656	.512	.966
Aphid Population Levels					
APOP 1	.565	-.757	.893	-.302	.579
APOP 2	.400	.411	.118	.483	.391
APOP 3	.226	.520	-.129	.430	.224
APOP 4	.124	.039	-.041	-.135	.132
APOP 5	.116	.825	-.341	.769	.096
Environmental Factors					
Elevation	.588	-.490	.979	.122	
Slope	-.183	-.016	-.223	-.184	
Aspect	-.035	.328	-.201	.293	
Disturbance	-.201	.301	-.129	-.532	
Geo. Area	-.212	-.652	.018	-.863	
Community Factors					
Percent Fir	.708	-.137			.713
Percent Spruce	-.509	.199			-.513
Stand Density	.631	.119			.631
Mean DBH Fir	-.128	.255			-.132
Mean DBH Spruce	-.204	.072			-.204
Mean DBH All	-.241	.224			-.245

57

TABLE 4-2

Canonical Correlation Coefficients for Balsam Woolly
Aphid Populations (1976) Throughout the Spruce-fir
Type in the Great Smoky Mountains (84 plots).

CANONICAL VARIABLE	ALL FACTORS		ENVIRONMENTAL FACTORS ONLY		COMMUNITY FACTORS ONLY
	#1	#2	#1	#2	#1
CANONICAL R	.974	.619	.639	.439	.869
Aphid Population Levels					
APOP 1	.658	-.618	.912	.140	.653
APOP 2	.371	.348	.065	.364	.411
APOP 3	.202	.104	.007	-.080	.208
APOP 5	-.132	.912	-.765	.579	-.157
Environmental Factors					
Elevation	.507	-.558	.941	.226	
Slope	-.025	.110	-.107	.146	
Aspect	.115	.295	-.094	.503	
Disturbance	-.184	-.199	-.077	-.582	
Geo. Area	-.004	-.541	.354	-.466	
Community Factors					
Percent Fir	.628	-.242			.669
Percent Spruce	-.357	.376			-.469
Stand Density	.684	.265			.751
Mean DBH Fir	-.225	.105			-.252
Mean DBH Spruce	-.055	-.079			-.064
Mean DBH All	-.330	.060			-.365

58

TABLE 4-3

Canonical Correlations Coefficients for Balsam Woolly
Aphid Populations (1977) Throughout the Spruce-fir
Type in the Great Smoky Mountains (47 plots).

CANONICAL VARIABLE	ALL FACTORS		ENVIRONMENTAL FACTORS ONLY		COMMUNITY FACTORS ONLY
	#1	#2	#1	#2	#1
CANONICAL R	.979	.680	.704	.660	.971
Aphid Population Levels					
APOP 1	.382	.858	.771	-.536	.471
APOP 2	.420	-.538	-.011	.451	.343
APOP 3	.328	-.697	-.055	.725	.262
APOP 4	.172	-.237	-.278	-.036	.141
APOP 5	.377	-.644	.039	.707	.321
Environmental Factors					
Elevation	.529	.557	.962	-.114	
Slope	-.265	.002	-.422	-.279	
Aspect	-.263	-.484	-.431	.415	
Disturbance	-.363	.542	-.321	-.830	
Geo. Area	-.451	.448	-.447	-.759	
Community Factors					
Percent Fir	.710	.229			.730
Percent Spruce	-.599	-.186			-.616
Stand Density	.627	-.115			.620
Mean DBH Fir	-.019	-.299			-.033
Mean DBH Spruce	-.310	-.216			-.330
Mean DBH All	-.192	-.273			-.204

59

TABLE 4-4

Canonical Correlation Coefficients for Balsam Woolly
Aphid Population Levels (1976 plus 1977) in the Western
Geographic Areas of the Great Smoky Mountains (81 plots).

CANONICAL VARIABLE	ALL FACTORS		ENVIRONMENTAL FACTORS ONLY		COMMUNITY FACTORS ONLY
	#1	#2	#1	#2	#1
CANONICAL R	.959	.683	.729		.956
Aphid Population Levels					
APOP 1	.687	-.704	.981		.688
APOP 2	.127	.679	-.396		.123
APOP 3	.025	.638	-.462		.015
APOP 4	.218	.285	-.082		.218
APOP 5	-.001	.709	-.537		.004
Environmental Factors					
Elevation	.495	-.769	.972		
Slope	-.205	.231	-.348		
Aspect	-.188	.345	-.411		
Disturbance			.012		
Community Factors					
Percent Fir	.624	-.154			.626
Percent Spruce	-.395	.282			-.397
Stand Density	.577	.168			.579
Mean DBH Fir	-.156	.288			-.156
Mean DBH Spruce	-.091	.188			-.092
Mean DBH All	-.247	.271			-.247

60

2. Typical <u>slightly infested</u> stand was based
    on a medium to strong positive APOP-1 correla-
    tion coefficient with the other APOP levels
    moderately to weakly, directly correlated.

3. The typical <u>moderately infested</u> stand was based
    on either a strong positive APOP-2 or APOP-3
    correlation coefficient in association with
    weak APOP-5 and APOP-1 correlation coefficients.

4. The typical <u>heavily infested</u> stand was based
    on very strong positive APOP-5 correlation
    coefficient with APOP-2, APOP-3, and APOP-4
    directly correlated and APOP-1 strongly negatively
    correlated.

Similar trends occurred in all analyses comparing aphid population levels and community-environmental factors. Therefore
"typical" communities are described with confidence in their real-
ity, although no such community existed. The most significant
characteristic is presented first followed in decreasing order
to the lease.

Typical Uninfested Stand

The typically uninfested stand was identified in 6 of 12 analyses and it was best characterized according to these community-
environmental factors.

<u>Elevation</u>. Uninfested stands were strongly correlated with the

higher elevations of spruce-fir type within the Great Smokies
(above 1768 meters).

Percent Fraser fir. Uninfested stands were associated with
communities containing at least 60 percent Fraser fir.

Stand density. Communities with a high stand density were
associated with uninfested stands. Fraser fir is capable of
growing in very dense communities.

Aspect. Eastern aspects (NE, E, and SE) were more likely to
support uninfested stands than other aspects.

Geographic area. Uninfested stands were moderately correlated
with the western portion of the spruce-fir type in the Park. There
were no uninfested stands in the eastern portion of the Park, characterized by the Pecks Corners area and Mt. Guyot. There were
some uninfested stands on Mt. LeConte, Mt. Kephart, Porters Mountain,
and Clingmans Dome.

Percent red spruce. Uninfested stands did not contain more
than 25 percent red spruce.

Slope. Slopes below 35 percent grade were more likely to support uninfested stands than steeper slopes.

Mean diameter all trees. Regardless of species composition,
uninfested stands were more likely to be comprised of trees less
than 15 centimeters in diameter than larger trees.

Fraser fir diameter. Fir communities comprised of trees less
than 14 centimeters in diameter were more likely to be uninfested

than infested. Stands composed of larger Fraser fir experienced
higher aphid population levels.

Typical Slightly Infested Stand

Slightly infested stands were identified by 4 of the 12 canonical correlation analyses. The trends were similar to that previously discussed with uninfested stands.

Stand density. Slightly infested stands were strongly correlated
with high stand density, as expressed by the sum of diameter for all
trees on each plot. Values ranged from 95 to 621 total centimeters
of diameter; the mean was 287 centimeters per plot. Stands of high
density were characterized by more than 300 total centimeters of
diameter per 100 square meters of ground surface.

Percent Fraser fir. Slightly infested stands in addition to
being dense, were comprised of greater than 60 percent Fraser fir.

Elevation. Slightly infested stands were only moderately
correlated with the high elevations (greater than 1768 meters) in
the spruce-fir type. Progressively lower elevations revealed
higher percentages of balsam woolly aphids.

Percent red spruce. There was some indication that slightly
infested stands were comprised of less than 25 percent red spruce.
This trend was not strongly developed, however.

Diameter all trees. There was only a weak correlation between
slightly infested stands and those containing trees less than 15

centimeters in diameter, regardless of species composition.

Fraser fir diameter. In contrast to uninfested stands, there was only a weak correlation between slightly infested stands and Fraser fir communities comprised of fir less than 15 centimeters in diameter.

Typical Moderately Infested Stands

Combinations of aphid population levels and community-environmental factors that would identify typical moderately infested stands were not statistically significant in any of the canonical correlation analyses. Our system of field coding individual trees and plots probably made this category difficult to isolate in analysis. Aphid behavior and aphid/host interaction are such that a moderately infested stand is short lived in the field.

Typical Heavily Infested Stand

Typical heavily infested stands were identified by six analyses. The following characteristics were associated with these stands.

Geographic area. Heavily infested stands were strongly correlated with geographic area. Communities located within the eastern geographic areas: Mt. Sterling, Big Cataloochee, and Mt. Guyot, were heavily infested and predominantly comprised of dead fir and a few live spruce. Those stands located in the western portion of the Park were not heavily infested. These patterns were described previously.

64

Elevation. Heavily infested stands were strongly correlated with elevations less than 1768 meters. Only in those geographic areas exposed to the aphid for the longest periods, were stands heavily infested at high elevations. Undoubtedly this reflects future trends in the remaining stands, however.

Aspect. Heavily infested stands were more likely to be found on western to northern aspects than on any other. Uninfested stands were on eastern and southeastern aspects.

Occurrence of disturbance. Heavily infested stands were moderately correlated with areas that had experienced some form of disturbance, predominantly logging, fire, or extensive windfall.

Fraser fir diameter. Those communities containing large fir trees (greater than 15 centimeters) were slightly correlated with heavily infested stands. These results substantiate those found in a related study reported in the next section.

Diameter of all trees. Regardless of species composition heavily infested stands were more likely to contain large trees (larger than 15 centimeters diameter) than uninfested stands. Large trees are excellent interceptors of air-borne aphids.

Percent red spruce. Heavily infested stands were weakly correlated with communities containing a high percentage of spruce (greater than 40 percent).

Slope. Slopes greater than 50 percent were weakly correlated with heavily infested stands. Uninfested stands occurred

on lesser slopes and at higher elevations, i.e., along ridges and
summits.

Summary of Community-Environmental Results

Several trends developed when the five balsam woolly aphid
population levels were categorized by community-environmental
factors. For example, elevation was important in all cases.
There was a lack of aphids at high elevations, slightly more
aphids were evident at slightly less elevations, and at the lowest
elevations the balsam woolly aphid population had maximized. Aspect, stand density, proportion of Fraser fir in contrast to red
spruce, and size of trees were some other characteristics that were
correlated with aphid abundance.

Elevation. The strongest realization in this study was that
aphid populations were least at the highest elevations and greatest at the lowest elevations of Fraser fir. Balsam woolly aphid
infestations started at the lowest elevations and progressed upslope. See Figure 4-3. Large Fraser fir on Mt. Sterling (1780
meters) have all been killed by the aphid, which started near the
hardwood-fir ecotone. Similarly the aphid has killed most of the
fir on Big Cataloochee and Mt. Guyot where only some remnant
Fraser fir stands remain at the highest elevations. Infested areas
around Mt. LeConte and Clingmans Dome now confirm these trends.
The newest and least well-developed infestations in these western
geographic areas were at low elevations, providing clear evidence

that progression will be toward the top! The infestations in the
western geographic areas reflect the length of exposure to the
balsam woolly aphid as well as the elevational gradient development.

Geographic area. That there was a correlation between aphid
population levels and geographic area was also apparent. The balsam
woolly aphid first came into the Park on the northwest slope of Mt.
Sterling. There were no Fraser fir trees any closer to Mt. Mitchell
within the Park. In ensuing years the aphid spread at variable
rates to include the entire spruce-fir type. At the time of this
study, infestation development was intermediate between the eastern
and western portions of the Fraser fir type. There was no evidence
that trees in the western geographic areas were less susceptible to
the aphid than trees in the eastern areas.

Community structure. Community structure has also been identified as influencing aphid infestations. Those dense stands of
small diameter Fraser fir are typical of most high elevation stands
in the Great Smoky Mountains. There was evidence here and elsewhere (Oosting and Billings, 1951) that red spruce is not abundantly
present at high elevations. Lack of balsam woolly aphid populations
in these high elevation stands appeared correlated with possible
time of exposure at that elevation. Aphids had not had enough
time to fully colonize these stands. However, there was some evidence that dense, young stands of Fraser fir above 1829 meters

67

Figure 4-3. Balsam woolly aphid infestations started at low elevation and progressed upslope. This ridge near Icewater Springs still had healthy fir along the ridge top but there were dying trees at midslope and dead ones along the creek.

elevation may be least susceptible to balsam woolly aphid attack. (Figure 4-4).

The only peak in the Park that has been subjected to balsam woolly aphid attacks sufficiently long to warrant such statements is Big Cataloochee. There were three disjunct patches of Fraser fir near the summit (1866 meters), two of which were young, dense, and apparently free of aphids. The third stand was older, had larger trees of differing crown and bark characteristics and the woolly aphid was more abundant. Undoubtedly the latter stand will be the first to succumb, but all three of them have all survived fifteen or more years of constant balsam woolly aphid presence. The larger dominant and codominant Fraser fir in the surrounding stands have long been dead, substantiating the revelation that stands of larger fir are first and most easily killed.

IMPACT OF THE APHID ON INDIVIDUAL FRASER FIR

In addition to thorough sampling of community-environmental factors associated with balsam woolly aphid infestations, individual trees were studied to identify those characteristics that most typified trees of high susceptibility to the balsam woolly aphid. It was not an objective to search and identify those trees exhibiting resistance to the aphid. However, that such trees exist is a fervent hope. It remained our goal to identify the impact on individual trees imposed by attack of the balsam woolly aphid.

70

Figure 4-4. This stand near the summit of Big Cataloochee
(1866 meters) was apparently free of aphids, in spite of a
considerable time of exposure to them. From the air, the
canopy presented an extremely dense and uniform surface;
no tree extended above the canopy to screen aphids out of
the winds.

Crown Color and Aphid Population

Significant relationships were found between crown color and aphid population (Chi square = 87.57 with probability of greater value under H_0 = 0.0001). Trends were as follows.

(1) A majority (85.7 percent) of the trees with green crowns was infested; conversely a majority (76.3 percent) of the uninfested trees had green crowns. Some "uninfested" trees had already reached the advanced stages of decline.

(2) Of the trees supporting the heaviest aphid population, 40 percent were green and 53 percent were light-fading.

(3) Light-fading trees usually had intermediate aphid population levels.

(4) By the time trees had reached the late stages of decline, the aphid population was reduced: 63.2 percent of the red-dying trees and 72.4 percent of the brown-dead trees did not have aphids.

Tree Height

Height and aphid population. Height did not differ significantly among trees having different levels of aphid population on the bole. However, severely infested trees were tallest (14.3 meters) and moderately infested ones were shortest (10.3 meters). Trees slightly infested or uninfested were of intermediate height. This trend was most pronounced in the Mt. Kephart and Clingmans Dome areas; elsewhere heights varied by only a few meters.

73

Height and crown color. There were significant differences between the following comparisons:

(1) height of green and height of brown dead,

(2) height of green and height of red-dying,

(3) height of light-fading and height of brown-dead

(4) height of light-fading and height of red-dying, and

(5) height of red-dying and height of brown-dead.

Green and light-fading crowns had the same mean height.

Height was greatest for red-dying trees (13.6 meters) and least for the brown-dead ones (10.3 meters). Trees with green or light-fading foliage had intermediate height (11.9 meters). These trends were most pronounced in the Mt. Kephart and Clingmans Dome geographic areas.

Diameter

Diameter and aphid population. Diameter was significantly different between trees that were severely infested and those that were moderately infested. The most severely infested trees had the largest diameter (19.7 centimeters), and those moderately infested had the smallest (15.0 centimeters). This was most pronounced in the Clingmans Dome area.

Diameter and crown color. The brown-dead trees had the smallest diameter (15.8 centimeters), and the red-dying ones had the largest (22.5 centimeters). Other comparisons of diameter and crown color were not significantly different. Trees with green and light-fading

crowns had intermediate diameter. These trends were most pronounced in the Hughes Ridge, Clingmans Dome and Mt. Kephart geographic areas.

Crown Class

Crown class and aphid population. Aphid population levels did not differ significantly with crown class (Chi sq. = 87.57, probability of a greater value under H_0 = 0.2209). The following trends were observed, however

(1) crown class distribution was fairly even among the uninfested trees,

(2) a majority (40 percent) of the slightly infested trees were intermediates,

(3) a majority (38.5 percent) of the moderately infested trees were suppressed, and

(4) among the severely infested trees, none were suppressed and there was an equal percentage (33.3 percent) of dominants, codominants and intermediates.

Crown class and crown color. The relationship between crown class and crown color was not statistically significant (Chi sq. = 3.813, probability of a greater value under H_0 = 0.9866). However, the following trends were observed:

(1) crown class distribution was fairly uniform among trees with green crowns, and

(2) of the trees in the most advanced stage of decline (those

with brown-dead crowns), 37.9 percent were suppressed and 13.8 percent were dominants.

Age

Sample tree ages ranged from 15 to 166 years; there were no significant differences in ages among trees in the various aphid population and crown color categories.

Bark Characteristics

Bark thickness. Differences in bark thickness among trees having different levels of aphid population and crown color were not significant. Bark thickness ranged from 1 millimeter to 14 millimeters.

Bark texture. Bark texture was not correlated with either aphid population or crown color (respectively: Chi sq. = 7.70, probability of greater value under H_0 = 0.5645, and Chi sq. = 6.62, probability of greater value under H_0 = 0.676).

Bark epiphytes. Bark epiphytes were correlated with both aphid population and crown color (respectively: Chi sq. = 19.05, probability of greater value under H_0 = 0.0248, and Chi sq. = 21.05, probability of greater value under H_0 = 0.0124). The trend was for highest aphid populations where bark epiphytic cover was light to moderate. Aphid damage as indicated by crown color was heaviest among trees with light epiphyte cover.

Growth Rate and Aphid Population

Growth rates did not differ significantly among trees having

different levels of aphid population. The following trends were observed, however:

(1) In 1976, the greatest annual radial increment was found among trees with no aphids (0.76 millimeters/year), followed by trees with moderate infestations (0.51 millimeters/year).

(2) When average growth rate was computed for the past 5, 10, 15, and 20 years, the most rapid growth was found in the uninfested trees, while those with the heaviest populations had the slowest growth.

Growth Rate and Crown Color

Radial growth rates among trees having different crown color were not significantly different for any of the time periods computed. However, the following trends were observed

(1) Radial growth rates for 1976 were greatest among the light-fading and green trees (0.76 millimeters/year) and least for the red-fading and brown-dead ones (0.51 millimeters/year).

(2) Radial growth rates for the last 5, 10, 15, and 20 years were generally most rapid for the green trees.

CHAPTER V

THE GREAT SMOKY MOUNTAINS
WITHOUT FRASER FIR?

The spruce-fir forest type constitutes one of the most important forest resources in the Southern Appalachians, but not for the economic worth of the timber. These last acres of the original 0.6 million hectares of spruce-fir have withstood logging, fires, grazing, farming, plus other uses and abuses. Although greatly reduced with only 20,000 hectares left in 1960, the resource was still intact. Due to market changes, technology changes, and logging damages, the forest is not valuable to the logger anymore. Spruce is no longer the highly desired species that it once was for paper manufacture. The principal products from these forests today are "services" not goods!

Recreation in many of its varied forms is the foremost service derived from these forests. There have been more yearly visits to the Great Smoky Mountains National Park that to any other. Currently, more commercial recreational services are available than were imaginable ten years ago. There is pressure to create wilderness status for the resource in Congress, in the

states, and in the region. The intensity of backpacking has
grown such that reservations are now necessary even at a primitive
campsite. Only the resident black bears are excused from this re-
quirement.

Significant alteration of the spruce-fir forest by any means
will have a dramatic effect upon visitors, especially on those
peaks of maximum visibility, e.g., Mt. LeConte, Clingmans Dome,
and along the Dome Road. Death of Fraser fir in these high elevations would have an immeasurable effect on socio-economic uses of
the area. What would happen to much of the backcountry use?
Would visitor-use patterns change enough to affect concessionaires
in the periphery of the Park? Would the loss of a portion of the
high elevation scenic resource put more visitor pressure on other
areas within the Park? Clearly the impact of losing these forests
would be major. Unfortunately these values are more intangible
than direct, consequently speculation is more abundant than fact.

APHID DISTRIBUTION IN THE SOUTHERN APPALACHIANS

Balsam woolly aphids were first confirmed in the Southern
Appalachians on Mt. Mitchell, North Carolina, in 1957 (McCambridge,
1958). The entire 3000 hectares of Fraser fir type were infested
at the time of discovery (Nagel, 1959), indicating a long history
of balsam woolly aphids on Mt. Mitchell.

Balsam woolly aphids could not have chosen to invade a better

peak in order to further their population expansion. Mt. Mitchell, the tallest peak in the East is centrally located to all Fraser fir in the Southern Appalachians. The Black Mountains, of which Mt. Mitchell is a part, have a north-south orientation in an otherwise southwest-northeast oriented chain. Therefore, Mt. Mitchell is a distinctively tall peak, rather than just another peak in a continuum of tall peaks. This geography when combined with wind movement for maximum geographic effect, insured abundant aphid dissemination throughout the region.

Balsam woolly aphids were detected in 1962 on Roan Mountain (Ciesla and Buchanan, 1962), which is located 32 kilometers N 15^0 E of Mt. Mitchell. Grandfather Mountain was found to have aphids in 1963; it is 48 kilometers N 50^0 E from Mt. Mitchell. The same year, aphids were found on Mt. Sterling which is 72 kilometers S 85^0 W of Mt. Mitchell (Ciesla et al, 1963). In subsequent years balsam woolly aphids have been found in all Fraser fir associations except the one on Mt. Rogers in Virginia.

Wind appeared to be the most probable vector of aphid dis-simination (Amman, 1966). Wind velocities and directions were adequate during motile periods of balsam woolly aphid life cycles to infest the new areas during the epidemic on Mt. Mitchell. Weather records showed at least one period with 56 kilometers per hour winds heading from Mt. Mitchell toward Mt. Sterling during the suspected time of infestation. Such winds were sufficient to

move aphid eggs and crawlers for the distances involved.

Within the Park

The early years. The initial infestation within the Great Smoky Mountains National Park was on Mt. Sterling Ridge, about 1.6 kilometers north of the summit. When discovered, the infestation covered an area approximating 600 meters long by 150 meters wide. There were 45 trees already dead or dying (Ciesla et al, 1963). Rotholz analyses of the dying trees revealed that the aphid had been present, albeit undetected, prior to 1958. The exact attack date could not be confirmed.

The National Park Service did make an attempt to contain the aphid within the original area by cutting infested trees; since the aphid is an obligate parasite, death of the tree will cause death of the aphids. The aphids were either more widely distributed than originally hoped before cutting started or else the agitation from cutting distributed them even further. Containment efforts were discontinued when it was realized that failure was imminent.

During 1964, 7 separate, small infestations containing 50 trees with stem infestations were found south of Butt Mountain, which is a spur ridge leading southeast from Big Cataloochee (Ciesla et al, 1964). There were other stands of Fraser fir between Butt Mountain and Mt. Sterling, but their distribution was not contiguous. Fires and logging during the early Twentieth

81

Century on both sides of Mt. Sterling Ridge created patchy patterns
of Fraser fir in this area. It appeared that the aphid had jumped
some of the Fraser fir stands near Pretty Hollow Gap between these
two mountains. This was an early indication of how the insect was
going to encircle much of the fir type before merging adjoining infestations. There was not the need to spread from one stand into another along a continuum. These leapfrog dissemination patterns would speed aphid population expansion within the Great
Smoky Mountains National Park.

By 1965, balsam woolly aphids had colonized the entire northwest slope of Mt. Sterling and some had reached the summit (1780
meters elevation). It would take several years to kill all the fir
on the summit. New mortality was not detected within the Butt
Mountain infestations, but the number of living, infested trees
increased. During 1965 no new areas of infestation were observed
(Lambert and Ciesla, 1966).

Aphids were detected on the southeastern slopes of Mt. Guyot,
at the headwaters of Big Creek in 1966. This first Mt. Guyot
infestation was located at the northern hardwood-Fraser fir ecotone,
an occurrence that was to be repeated many times throughout the
Park. Also during 1966 the entire Mt. Sterling Ridge became infested when separate infestations merged around Big Cataloochee
and Mt. Sterling (Lambert and Ciesla, 1967). Three patches of fir
on the summit of Big Cataloochee remained aphid free, a condition

which still existed in 1977.

During the next several years, the aphid continued to expand its population. Although 1967 was an uneventful year, during 1968 the aphid colonized the northern portion of the disjunct spruce-fir type in the Heintooga area. When dying trees were first spotted in the Heintooga Geographic Area, approximately 80 hectares on Spruce Mountain were already infested. Now the Balsam Mountains were infested all the way from Big Cataloochee to Polls Gap. The aphid was also occurring in the northern Balsam Mountains where a new infestation was found on a ridge leading north from Luftee Knob.

The explosion. Events during 1969 would prove to be a harbinger of what was to come. Mortality on Mt. Sterling was far more severe than at any time or place in the history of the aphid in the Park; the upper slopes and the summit of Mt. Sterling were completely covered with thousands of red-fading Fraser firs. After encircling the summit during the preceding 11 years, the balsam woolly aphid population exploded bringing complete destruction to fir on Mt. Sterling.

Cognizance should be taken that this process has not yet been repeated on any other summit to date. Although Big Cataloochee and Mt. Guyot were found to be infested soon after Mt. Sterling, three patches of fir remain alive on Big Cataloochee and the summit of Mt. Guyot is still clothed in fir. There were aphids on the summit of

83

Big Cataloochee, for mature fir had died all around the healthy
patches. A few aphids were also found near the benchmark (2018
meters) on Mt. Guyot, but most mortality had not occurred closer
to the summit than 1966 meters elevation.

Aphid population development on Big Cataloochee appeared intermediate between that on Mt. Sterling and Mt. Guyot. The overstory fir on Big Cataloochee were all dead along the low- to mid-
slopes, including the prominent spur ridges. Because logging had
extended to the summit out of the Cataloochee Valley, these fir
stands were remnant veterans or advanced reproduction that had
reached pole-size before aphid mortality became severe. Tree
sizes were similar to those on Mt. Sterling and southeastern slopes
of Mt. Guyot where logging and fires had created comparable stands
and the aphid had also caused mortality. However, on the summit of
Big Cataloochee there existed a dense, young stand of Fraser fir.
There was a similar stand toward Big Butt, both of which probably
originated well after logging had ceased. A third stand toward
Balsam Corner was older and it probably was too small during the
logging era to attract attention. Although still quite dense, this
latter stand had now reached pole-size.

Aphid presence in the Big Cataloochee stands varied, not according to elevation, for the summit is relatively large and flat, but
rather according to tree size and stand density. The summit stand
was free of aphids, as per our observation techniques. The Big
Butt stand had a few aphids on some of the larger trees, but they
had to be climbed to find them. Neither of these stands had aphid-

caused mortality. The Balsam Corner stand had aphids and some of the crowns were fading. Clearly the Balsam Corner stand will be the first to experience maximum mortality and probably soon. The future of the young, dense stands is speculation at this point.

Fraser fir stands on Mt. Guyot have not been influenced by logging as much as those on Big Cataloochee and Mt. Sterling. Only the southeastern slopes were logged and probably not all the way to the summit. The northwest slopes contained some of the largest Fraser fir and red spruce observed during the entire project, clearly reflecting a lack of logging. When the aphid population explodes across the summit of Mt. Guyot, the color response should parallel that witnessed on Mt. Sterling in 1969; tree size and stand density are comparable to Mt. Sterling instead of Big Cataloochee.

Another difference between these peaks is their elevation: Mt. Sterling is 1780 meters elevation, Big Cataloochee is 1866 meters elevation, and Mt. Guyot is 2018 meters elevation. There is evidence on other peaks that aphid population development slowed as high elevation was reached. However, there is no reason to speculate that high elevation will preclude aphid population expansion to any summit in the Great Smoky Mountains National Park; Mt. Mitchell at 2037 meters elevation was completely colonized by the balsam woolly aphid, but after an undetermined period.

The only new infestation detected during 1969 was on the southeastern slope of Luftee Knob at the headwaters of Dans Creek,

a tributary of Straight Fork.

Activity During the Seventies

Aerial detection within the National Park was not maintained
regularly by the U. S. Forest Service after 1970. Aphid population
expansion was not chronicled as thoroughly as it might have been.
National Park Service personnel were not maintaining any control
programs because there were none available, and except for the color
display on Mt. Sterling the infestations were too scattered and
small to attract much attention from backcountry visitors.

During the early seventies just a few new infestations were located. Existing hotspots grew larger and mortality occurred, sometimes alarmingly. During 1972 the northern section of the Balsam
Mountains between Big Cataloochee and Tricorner Knob experienced
severe mortality at the lower elevations of fir. Fir trees on the
summit of Luftee Knob and along the main ridge of the Balsam Mountains
were not killed at this time. Severe fir mortality was also found
in several sections of the Heintooga Geographic Area, especially
around Spruce Mountain and Cataloochee Balsam.

It was soon realized that aerial detection was limited to
revealing trees that were dead or dying, even with color infrared
film. Due to the inconspicuous size and complex life cycle of
the balsam woolly aphid plus the delayed action of their infestation
development, ground checks were devised by the U. S. Forest Service.

86

Microscope slides coated with a sticky material were mounted on
tree stems. Waiting for an unsuspecting aphid looking for a
home, these traps were placed along the Appalachian Trail and
the Clingmans Dome Road. After extensive, daily maintenance for
a period, no balsam woolly aphids were found west of Tricorner
Knob. Aphids were well-documented east of Tricorner Knob all
the way to Mt. Sterling along Mt. Sterling Ridge, in the Heintooga
area, and in the Mt. Guyot section of the State-Line Ridge.
Except for an occasional fly-over and a few end-of-the-roll pic-
tures, surveillance was terminated after 1972.

Based on the information in our project, there was no pause
in the advance of aphid population in the Park. The infestations
along Forney Ridge, Noland Divide, Mingus Lead, and Mt. LeConte
have all been developing longer than four years. Using the May,
1976 photography as reference, the size of the above infestations
and the extent of dead trees they contain showed a history of 7
to 12 years to be more likely. How were these infestations missed
by the extensive trapping system?

In every case where records existed for the initial locus for
a developing infestation, it occurred near the northern hardwood-
fir ecotone. On most ridges this was not an easy place to reach.
There were few trails, certainly not the Appalachian Trail, that
went into these hardwood-fir ecotones. Because the trapping system was monitored daily, speed was important; there wasn't time

to do much off-trail work. Secondly, the traps were placed at high elevations, along the trails or the Dome Road. Only by 1977 have the aphids come close to the State-Line Ridge in some locations west of Tricorner Knob.

There is some credence in the idea that balsam woolly aphids moved directly from Tricorner Knob and Mt. Guyot into the Mt. LeConte and Clingmans Dome areas. Infestations there were as well-developed, if not moreso, than those between Tricorner Knob and Newfound Gap. This middle portion of the spruce-fir distribution does not have the celebrated elevation of Guyot or LeConte or the Dome, and it also lacks the Fraser fir of these peaks. Settlement and logging at low- to mid-elevations plus the inevitable fires have greatly shaped the plant communities around Charlies Bunion and Porters Mountain. Hardwoods have replaced spruce-fir in many areas, as witness numerous aerial photos.

Summary. Balsam woolly aphids have expanded their population throughout the entire Great Smoky Mountains National Park. Infestations have developed to near maximum fir mortality east of Tricorner Knob. Most living Fraser fir in those areas are saplings or advanced reproduction seedlings, some of which have grown with increased vigor upon release from the overstory. The scenic resource has been lost in these areas, hopefully just temporarily.

Aphids have killed Fraser fir along all trails leading to Mt. LeConte. As the many infestations merge, the impact will be more

painfully obvious. Already large firs around the Lodge are infested.
As these begin to die and the aphid population explodes all over
the summit, the coloration on Mt. LeConte will be breath-taking,
unfortunately not in the usual sense. For the last few years,
some trees have been dying along the Clingmans Dome Road. The pace
will quicken when the Noland Divide, Big Slick Ridge, Sugarlands
Spur, and Forney Ridge infestations reach the road on their way
to the State-Line Ridge.

APHID IMPACT ON FRASER FIR COMMUNITIES

It's time to recapitulate a few facts. Uninfested stands
were mostly found at the highest elevations possible. Heavily
infested stands were found at lower elevations, and there was a
definite gradient of aphid population levels between these two extremes.

Another fact: all known aphid infestations began at low
elevations along the northern hardwood-fir ecotone and spread upslope. Figure 4-1 clearly showed the intermediate development of
infestations on the slopes of Mt. LeConte and Clingmans Dome, substantiating this fact.

Since we were lacking training in aphid psychology and motive
preferences, it was not immediately clear to us why aphid infestation patterns assumed the shape that they did. There were no overwhelming explanations, however, our research has offered some

plausible inferences.

There are some well-documented ecological differences that occur along elevation gradients in terms of community structure,
species composition, size of trees, and site productivity. There
are also some less obvious factors to consider. For example,
our extensive field exposure has caused us to notice wind speeds
and wind behavior. Perhaps because the aphid is passively dis-
seminated, its preference for hosts is determined by random chance
and by physical reasons. Unfortunately, many of the apparent
factors appear related, thereby causing some confusion as to which
was the cause and which was the effect. Being cognizant of the
dangers of discussion in a somewhat circuitous route, a number of
the more pertinent factors have been elaborated upon.

Wind Dissemination

Entomologists have written that balsam woolly aphids are not
capable of active dissemination, i.e., they are passively dissemin-
ated by forces other than of their own determination. Lacking
wings, they are blown about as eggs and in the crawler stages. They
are not selective in the choice of hosts available to them: if it's
a suitable attack site the aphid may colonize, otherwise it probably
dies unless wind moves it along again. There was no evidence that
these insects function actively in host selection, in contrast to
some other well-known forest insect pests.

The opportunity for us to spend considerable time in the spruce-

fir forest type and for extended periods has created an appreciation for how wind velocity and direction are affected both by
convection and topography. The bite of cold wind along an exposed
ridge or summit accentuates the effect of wind speed. Clouds laz-
ily drifting toward a ridge by convection gain speed as they ap-
proach the top. Suddenly, they explode across the top and disappear
on the leeward side. Watching the mist of early cloud formation
rush through a gap and spill down the leeward side with diminishing
speed, only to curl around in a giant eddy until the warm air at
lower elevation absorbs their visibility, illustrates clearly the
effect of wind rushing with accentuated speed through a restricted
pass and the formation of eddies. Given ample opportunity to sit
upon a rock after a long pull with more weight in the pack than was
comfortable, one begins to realize that there are balsam woolly aphids
in those winds!

Eddy Formation. When winds of some depth moving along the
earth surface encounter a physical obstruction, they rise to pass
over the mountain barrier. Because the air movement above the
mountain summit is relatively unaffected by the mountain, the ris-
ing air is restricted to less space; to accommodate the restriction,
the wind speed increases sharply as the highest barrier is reached.
The effect is similar to water in a broad, slow stream being restricted at the start of rapids. After the ridge or summit has been
passed by the speeding winds, air pressure is lower and air parcels

settle into these low-pressure areas on the leeward side. Meteor-
ologists write that as air spills onto the leeward slope
after crossing the summit, it starts to circle, thereby creating
an eddy. The stronger the wind, the larger the eddy, and the more
force within the eddy.

   Aphid movement. Rather than emphasizing meteorological prin-
ciples as the relevant discussion, it adds a new dimension to re-
consider wind patterns with balsam woolly aphid abundantly present
in them. Aphids riding the accelerating winds on the windward
slopes are not really in a position to select host firs; if by
chance they land in a tree they may not survive the collision!
However, as the eddy develops, the aphid-laden air slows and set-
tles into low pressure spaces. Balsam woolly aphids are more
gently deposited on trees at the low point of the eddy. If Fraser
fir are present, an infestation could develop on the leeward slopes,
maybe at the northern hardwood-fir ecotone where so many have started.

   Regardless of the speculative nature of how aphids get started
at low elevations, there is more substance to how the infestations
spread upslope. Large fir at low elevations are attacked and killed by the aphids first. As aphid populations increase in these
trees, surrounding firs are infested by any of various means and
the infestation grows into the typical hotspot witnessed on aerial
photos. Gentle, upslope convective winds during mid-day push aphids

into those next few prominent crowns above the hotspot. If red spruce intercepts aphids from the winds, subordinate firs may become infested, but if the large crowns are fir, the infestation is rapidly advanced. This pattern was quite apparent on Mt. Guyot during 1977.

On Mount Guyot. A secondary spur southwest of Mt. Guyot had a well-developed hotspot at mid-slope. Most of the fir were dead or dying in the hotspot center and scattered aphids were found close to the Appalachian Trail which was 100 meters above the hot-spot. Above the Appalachian Trail few aphids were detected. However, the plot at 1951 meters (above the Appalachian Trail) contained one large Fraser fir that was 7 meters taller than its neighbors. Growing beneath this dominant were two small Fraser firs, with one aphid on each stem. Convective winds brought the aphids from the hotspot toward the summit and the large fir intercepted some. Although it was still alive and growing, this dominant had become infested and it was infesting trees around it.

Summary. Winds very much affect the dissemination of balsam woolly aphids in mountainous regions. Strong winds and eddies bring aphids into new areas, winds facilitate spread at low elevations of Fraser fir where infestations start, and winds move aphids upslope to complete the colonization of available hosts on that mountain. Without wind, the balsam woolly aphid problem would probably lessen.

93

Host Responses to Aphids

When trees decline from balsam woolly aphid attack, crown
color first becomes light green with some leaves or occasionally
entire branches turning yellow to red; this procedure gradually
progresses from the bottom of the crown upward. Bright red
foliage throughout the crown follows and the foliage turns dark
reddish-brown as the tree dies. A severe bole attack can bring
about a complete color change sequence in a single season.
Trees with green crowns can be infested; there is a time lag of
a few years between aphid attack and crown color change.

The interval between initial infestation and subsequent
color changes varied with the duration and the severity of the infestation. There may also be individual host differences in resistance due to genetic or environmental circumstances that hasten
or delay the process of decline (Balch, 1952). Crowns thus provide
an indication of the amount of accumulated damage already sustained
by a tree plus they indicate the ability of a tree to provide an
environment conducive to woolly aphid establishment, survival, and
reproduction. The strong relationship between aphid population level
and crown color permitted their use as complementary indicators of
the intensity of balsam woolly aphid infestations.

Aphids prefer large trees. Tree size was the most important
factor in determining host suitability to the balsam woolly aphid.
There were consistent relationships between size characteristics,

e.g., height, diameter, and crown class, and aphid population levels.

Trees presently supporting the heaviest aphid populations were the dominants and codominants with the largest height and diameter. This relationship between tree size and aphid population was more distinct for height and diameter than for crown class, but by definition dominants and codominants have the largest diameter and height. There are several factors involved. The larger size of these trees increased their probability of intercepting air-borne aphids causing them to be the first in the stand to be infested. This additional time permitted the development of aphid populations larger than those on the smaller, adjacent trees. Secondly, vigorous large trees provided a favorable habitat for aphid population increases because of their increased metabolic efficiency and propensity to produce energy.

Within the Great Smoky Mountains National Park there were two separate but related facts concerning tree size and aphid population levels. Large trees supported the greatest aphid population levels, at least just before they died. These observations were most pronounced in the Mt. Kephart and Mt. LeConte Geographic Areas where infestation intensity and development were intermediate and correlations of tree size and aphid population levels have not been masked by a more fully developed infestation. Other areas lacked these trends because they had progressed too far or not far enough.

Secondly, elevation gradient transects revealed that the largest trees, regardless of species, were associated with the heavily

infested stands, and in most cases these trees were near the northern hardwood-fir ecotone. One clear observation was the lack of
many large trees at high elevations, thereby adding credence to
the hypothesis that tree size was correlated to aphid population
levels in a developing infestation. Balch (1952) and Johnson et
al (1963) observed that in newly infested areas, the insect first
became numerous on the large trees. In addition, Balch (1952)
suggested that high tree vigor favored the multiplication of the
insect. After an infestation had been active for several years,
woolly aphids became more generally distributed throughout the
stand, although there was still considerable variation between trees.

Trees with slight to moderate infestations had suppressed and
intermediate crowns and small stem diameters. Unfortunately,
there was no way to determine whether populations were increasing
or decreasing on these stems. The most probable explanation was
that intermediates had moderate but increasing populations while
suppressed trees had moderate but decreasing aphid populations.
This is likely, because crown color and tree size comparisons indicated that the most advanced decline was among the smallest trees.
The largest trees became infested first, and though aphid populations
were comparatively less on the intermediates, they were increasing.
Small, suppressed trees became infested about the same time as the
intermediates, but they succumbed rapidly and aphid populations
were decreasing on these dying trees suggesting that these small

trees may be poor hosts. However, a large population would not
be necessary to kill a tree that was already under stress.

Favorable habitat. There is limited information about the
types of host metabolic products that aphids require to sustain
life abundantly. It is possible that those trees at low elevations have evolved whatever physiological characteristics the
aphids need, while those at high elevations still lack them.
Our study was not designed to illuminate such differences. However, it is not likely that the limited elevation differences in
the Fraser fir distribution in the Great Smoky Mountains would
produce significant clinal variation in major metabolic functions.
The maximum elevation range was 805 meters, but the average was
closer to 375 meters before aphids even further restricted it by
killing the fir at low elevation.

It is much more plausible that whatever the aphids required
in a host's physiology, they got more of it from large trees which
were found at low elevation. Consequently, large trees built and
maintained large aphid populations before they died from the
parasites. One large, open-grown fir on Mt. Sterling Ridge near
the Mt. Sterling Gap Trail intersection tenaciously maintained
life and heavy aphid populations simultaneously. Its large crown
was producing enough energy for both host and parasite, albeit
the relationship cannot last much longer.

Better aphid habitat through morphological differences in
fir bark appear more likely than metabolic adaptations. Aphid

population levels were not correlated with bark texture because
there was so much variation in the sample that meaningful trends
were masked. Bark texture appeared so dependant on growth rate,
tree vigor, and stand density that direct correlations to aphid
population levels were not possible.

However, one cannot spend as much time in the woods as we
did and not develop some valid observations. Rarely, were balsam
woolly aphids present on saplings that had smooth bark unless an
isolated insect attacked near a lenticel or in the roughened area
below a branch. This relationship persisted as long as the tree
maintained smooth bark, up to 12 or 15 centimeters in diameter.
Some trees growing in an old clearing on Mt. Sterling were approaching seed bearing size and they retained smooth bark, characteristic of their rapid growth.

As trees developed in size or suffered reduced vigor due to
competition, their bark roughened and balsam woolly aphids found
more places to attack. Greenbank (1970) and Amman (1970) also
found that infestations were more severe on trees with rough bark
than those with smooth bark. They suggested that crevices and
lenticels provided protection for the insects. Fraser fir in a
natural stand undergoing normal competitive cycles will develop
roughened bark when they attain 12 to 15 centimeters in diameter.
After trees reached sufficient size, we did not witness bark texture precluding aphid population increases.

98

Stands on Big Cataloochee of extreme density experienced competition intense enough to affect growth and consequently bark texture. Most of those trees including some between 15 and 20 centimeters in diameter, did not have bark characteristics like other
Fraser firs of comparable size. Rather, the bark was reddish and
developed flaky fissures with tissue paper edges. Analyses were
not made of bark thickness or cellular morphology, but balsam woolly
aphids were conspicuously absent from these trees. One tree large
enough to climb and loaded with cones had three separate bark
textures along the stem. The red tissue paper bark was on the
lower stem, gray somewhat more normal bark was on the mid-stem,
and the smooth gray-green bark of young saplings was within the
mid-to upper-crown. Aphids were only found, and in limited numbers,
on the middle portion of the stem. There remains only limited doubt
that bark texture was not correlated to aphid population, if not
as surface physical resistance to attack, then perhaps as an
external realization of some internal modification that does not
facilitate aphid feeding.

Balch (1952) found that host resistance created by formation
of secondary periderm under affected tissues most often occurred
in areas where the bark was thick. He noted, however, that the
ability of the tree to provide itself with this protection depended upon its vigor and the portion of the tree that was heavily
infested. Bole infestations on Fraser fir were generally more

serious than the crown infestations of the balsam fir in Balch's study; this may explain why significant differences in bark thickness among the various aphid populations and crown color categories were not found. Woolly aphid stylets were long enough that outer bark within the normal range of thickness was not an important obstacle to penetration, and total bark thickness was not a determining factor.

Bark epiphytes (mosses and lichens) can provide protection from the weather for aphids; they may also protect the tree from aphids, although insects were sometimes found beneath moss and lichen cover. The motile larvae crawled between the bark and epiphytes. Our results indicated that the aphids became established and survived best on trees with a moderate cover of epiphytes; those with no epiphytes provided little protection and heavy cover prevented aphid attack.

Crown color and tree size. Separate analysis for aphid population levels associated with tree size and crown color have verified much of the results previously discussed. Suppressed trees suffered the heaviest damage, as indicated by crown color changes, and they were the first to succumb to aphid attack. These trees were apparently less able to withstand insect infestation because they were already under stress from competition and they were not vigorous enough to resist by building secondary periderm cells to protect the parenchyma. Bryant (1974) noted that this wound-healing

response was usually overcome during a continued heavy attack.

The tendency of short trees to be the most heavily damaged was more pronounced in areas of recent infestation. The effects of tree size differences and aphid preferences were masked as damage became more widespread in the stands. For instance, the larger trees that were red in recent infestations would soon become brown and increase the mean size of the dead group. This was apparent on Mt. Guyot where infestations were older. In a study of balsam woolly aphid infestations of Pacific silver fir in Washington, Johnson et al (1963) found the most severe damage on large trees; his work was done in areas where the insect had been established for some time.

In another study relating crown class to aphid damage, Schooley and Oldford (1974) found that in young stems of balsam fir in Canada, trees in the various crown classes had the same probability of receiving aphid damage. They were studying balsam fir in even-aged plantations. In our study of natural stands complete with varying age classes, aphid population levels showed differential preferences for crown class sizes.

Growth rate and aphid infestation. Previous studies have shown inter-relationships between aphid infestation and growth rate of the host tree; Balch (1952) and Amman (1970) suggested that vigorous, rapidly growing trees provided the most food for the woolly aphids and therefore supported the largest population. Both researchers

also observed a marked increase in annual increment during the first one or two years of infestation. According to Balch et al (1964), maximum growth stimulation occurred in vigorous trees with moderate infestation; heavy infestations tended to inhibit growth. They suggested that some substance in the aphid saliva was toxic in high concentrations and growth responses depended upon interaction with some factor in a tree that varied with its vigor. As infestations progressed, the physiological processes of the host were impaired and growth rate slowed before the tree died.

During 1976, growth rate was most rapid in the moderately infested Fraser firs and slowest in heavily infested ones. The rapid growth probably reflected the growth spurt associated with initial infestation while the heavily infested trees had begun to die and their growth had slowed. Since the infestations could not be accurately dated, there was no way to determine the relationships between current aphid population levels and growth rate prior to infestation. Average growth rates for the past 5, 10, 15, and 20 years were highly variable.

Trees with the largest annual increment had light-fading foliage. Their rapid growth rate could be in response to aphid attack, or a result of tree vigor, or a combination of these factors. Trees suffering the greatest damage in 1976 had grown most rapid during the last 20 years, confirming that rapidly growing trees became infested first, supported the heaviest aphid populations, and

experienced the growth spurt that accompanied initial attack.

Community Structure and Aphid Population

Spruce and fir are common community associates in those places in North America that have the proper environment to support them. Although the species composition is different in many of the regions, the communities have similar structure. The spruce component lives the longest, attains the greatest size, has the fewest seedlings to mature tree ratio, and has a history of highest economic value when compaired with the fir associate.

Spruce trees commonly live twice as long as fir in a natural, unmanaged stand; two fir life cycles are common while the spruce component matures into stately giants that are associated with the forest primeval. Fir has more rapid early growth rates than spruce, and as so commonly happens in tree communities, the early producers mature early. Because fir reproduction is quite shade tolerant, a second generation begins under the spruce. The new stratum will frequently remain subordinate to the spruce, except as spruce die and the understory is released to assume a position in the canopy.

Fir are prolific seed producers and germination is adequate following at least one winter of cold treatment. Fir seedlings frequently carpet the ground under mature canopies, even those with a significant percentage of spruce. However, few of them ever make

it into the canopy; seedling mortality is high unless release is
forthcoming. Spruce seedlings are less abundant but more of
them seem to grow into the canopy.

There were many natural and mancaused factors that shaped the
red spruce-Fraser fir stands in the Great Smoky Mountains. Stands
sampled in this project represented a variety of community structures: there were young, dense stands of pure fir; intermediate-
aged stands were characterized by large red spruce and a few scattered Fraser fir overtopping younger strata of reproduction, saplings, and poles. Occasionally the proportion of red spruce was
quite high, probably an indication of nearing the maturity of
the red spruce life cycle. Where original communities had been
disturbed by logging, current stand structure was not as predic-
table or as readily categorized as previously stated. The large
red spruce had been highly sought by loggers. Early loggers only
took Fraser fir if it was of exceptional quality or in the way.
Although stands recovered unless they were consumed by wildfire
each is uniquely its own representative in any scheme of classify-
ing overstory structure.

On the steep southeastern slope of Mt. Guyot where logging
progressed nearly to the summit, two different stands validate the
importance of man in shaping stand structure. The 1951 meter elevation plot occurred in a .stand of maturing Fraser fir and a few red
spruce; these were too small to warrant the effort to cut them when
the Big Creek drainage was logged. Down the slope was a stand of

nearly impenetrable Fraser fir reproduction about five meters
tall growing among the rocks exposed by fires and under a sparse
canopy of old but small red spruce. With such variability, analysis
of balsam woolly aphid preferences were not always as revelatory
as desired, leaving more to interpretation and description.

Stand density. Two of the community structural characteristics
associated with uninfested stands were high density and small stem
diameters. High density was related to tree size because the sum
of stem diameters was used as a density index. Numerous small
stems had a greater affect on the index than few large diameter
stems. Heavily infested stands were not dense nor were the trees
small, rather they were comprised of large trees at wide spac-
ings. High density was not correlated to elevation gradients.
This is significant, for it identifies stand density as a meaningful goal to minimize aphid impact.

At least two possibilities exist as to why aphids do not prefer
dense stands. From the air, dense stands present a tight, uniform
canopy; sometimes it was difficult to distinguish between rhododendron thickets and dense fir stands, especially if the latter were
young. The uniform canopy of a dense stand does not readily
provide many obstacles to intercept air-borne balsam woolly aphids.
They blow past, unless the wind forces them into the canopy or al-
lows gravity to pull them into the crowns. This structure was characteristic of young, dense, even-aged fir stands.

105

Dense stands with small diameter trees provide few places for aphids to attack the stems. The correlations of small trees and aphid population levels have already been discussed.

The combination of favorable environment and a prolific seed producing species frequently results in abundant reproduction. So it is with Fraser fir! Occasionally dense stands developed where the canopy had been disturbed and advanced reproduction had responded to release. In other instances the cause of disturbance was not evident, but the density and vigor of the young Fraser fir stand was just as intense. In those cases where Fraser fir was afforded an opportunity to be a pioneer species, dense young stands resulted regardless of elevation. These stands lacked favorable habitats for aphid population development.

In contrast, those stands of greater age, less density, and more irregular height profiles which were characteristic of uneven-aged conditions provided excellent opportunities for balsam woolly aphid colonization. There were plenty of crowns that had differentiated into dominants and they were readily intercepting aphids. In addition, their large size made possible aphid population increases because there were numerous attack sites and their metabolism was adequate to be a good host.

Cognizance should be taken that young, dense stands of fir and spruce were probably more common at high elevation than at lower elevations due to the types of disturbances that occur in natural,

unmanaged stands, i.e., blowdown, insect or disease, and fire.
Stands at high elevations are more frequently exposed to these disturbances than those occurring in more sheltered locations. Two
of the three stands of Big Cataloochee are examples of this fact.
Their origins which post-date the logging era were attributable
to windthrow, as exemplified by the accumulation of large, down
trees under the developing canopy.

In summary, aphid populations did not prefer young, dense
stands of trees, probably because there were less opportunities to
invade these communities due to uniform height profile and lack
of suitable attack sites on individual stands. Although not statistically correlated with elevation gradients, they appeared more
numerous at high elevations. Stands comprised of large trees at
less stand density characterized the lower elevations of the spruce-
fir type and they were more readily attacked by the aphids. Mortality was high.

Species composition. Stands which contained few aphids were
predominantly Fraser fir with correspondingly low percentages of
red spruce. In a separate analysis heavily infested stands contained high percentages of red spruce and low percentages of Fraser
fir.

Species composition had an important influence on aphid presence in the stand, but perhaps not in an obvious or anticipated
way. At no time did aphids colonize spruce trees, regardless of

their size, so why were heavily infested stands mostly spruce and
uninfested stands 60 percent Fraser fir? Elevation greatly affected species composition and we have already seen that elevation had
one of the most significant impacts on balsam woolly aphid distribution. Tall, mature spruce at low elevation in heavily infested
stands presented a substantial intercepting obstacle to air-borne
aphids. Even though the insect could not find an attack site on
the spruce, it was able to attack the subordinate or adjacent fir.

That aphid populations were highest in stands dominated by
red spruce was more a function of stand elevation and indirectly
species composition. The height of red spruce was also a factor
in presenting an irregular canopy. Pure stands of Fraser fir are
susceptible to aphids, but usually after the population has in-
tensified enough to attack the highest elevations. In the final
analysis, stands in which red spruce predominated will present the
least changed structure to the scenic-minded hiker.

Summary. Community structure has shown some important trends
in aphid population levels. Unfortunately, few of them can be interpreted as truly showing balsam woolly preferences for community
structure characteristics. Aphids were scarce in dense, pure, even-
aged, young stands at high elevation, at least in part because
they hadn't arrived in critical mass yet to attack with authority.
The apparent inability of the aphid to successfully build populations on small firs is really the only positive realization. Amman,

by personal correspondence, verified that seedling and saplings were infrequently killed by aphids unless they were already stressed. Regrettably, the extended summer drought of 1976 stressed thousands of seedlings and saplings to death on Mt. Guyot and undoubtedly on other peaks.

That aphid populations were correlated with not so dense, mixed, uneven-aged, mature stands of predominantly red spruce at low elevation was an accumulation of indirect occurrences. In those places where aphid population intensity was intermediate, we were able to discern the first stands to be attacked, but there was no evidence that aphids will be restricted to these stands. Rather they presented a combination of factors that allowed aphid populations to intensify, i.e., large trees and irregular height profiles. Unless individual trees can be found with some inherited resistance to aphid attack, all Fraser fir communities regardless of species composition, stand density, or other community factors will be seriously affected, if not eliminated.

Environmental Factors

This study was not designed for a detailed analysis of the environment enveloping Fraser fir communities. Such provisions would have required a separate study. However, some inferences were available from our project.

Aspect. Uninfested stands were more likely to be on eastern

(including northeast and southeast) aspects and infested stands
were correlated with western aspects (including northwest and southwest). Considering the random aphid dissemination mechanisms, it
was almost a surprise that such order was found in aphid populations.
Clearly, those mountains that had experienced aphid populations
for some time and had maximum population development did not con-
tribute much to this correlation, for aphids had completely sur-
rounded them on their way to the summit. However, those peaks
with new infestations did have more aphids on their western slopes
than anywhere else, e.g., the most severe infestations on Mt.
LeConte were approaching the summit from the west/northwest.

Realizing that passive dissemination greatly affected where
aphids attack plus the impossibility of validating disseminating
pathways in this study, there were still some inferences to be
gleaned. It is entirely possible that aphids landing on the western aspects, especially the southwestern aspects, found a more
favorable environment in which to develop an infestation. The
warm temperatures inherent on such aspects facilitate the speed
of metamorphosis, allowing maximum aphids per year. Greenback
(1970) and Atkins (1972) reported that warm temperatures shortened
the time from egg to adult, thereby maximum generations occurred.

Warm temperatures extended the growing season; those protected
slopes warm more quickly in the spring and they stay warm well into

the autumn. Balsam woolly aphids may have as many as four generations per year in the Great Smoky Mountains, and where is a more
likely spot than on a southwest to west aspect. An extended
growing season also means that host firs are going to provide more
of what balsam woolly aphids feed upon for a longer period.

Site quality. Previous studies have noted correlations in
aphid population levels and sited quality. Site characteristics
may have affected aphid populations directly by determining accessibility through aspect and elevation or indirectly through micro-
climate by influencing the growth rate and overall vigor of host
trees. Johnson et al (1963) found that aphid populations were usually high on sites with high site index for Pacific silver fir;
understory vegetation types were also correlated with site index.
Brower (1942) found in Maine that infestations on balsam fir were
most severe on those trees growing on poor sites with low-lying,
poorly drained soils. Balch (1952) could find no conclusive relationship between forest type, site quality, and degree of infestation
or damage on balsam fir. Bryant (1974) reported that damage was
more severe on dry sites than on moist sites in Newfoundland.

Johnson et al (1963) found that Oxalis was associated with
stands suffering severe infestations and Vaccinium with those having
the least damage. In our study, the two vegetation types most common under maximum infestations both contained Oxalis as a major component. The presence of such understory species reflects a combination

of environmental conditions more favorable to aphid population development, either affecting the insects directly of perhaps indirectly through the host tree.

Summary. Environmental stimuli greatly affect the development of balsam woolly aphid infestations. Although this project was not designed to illuminate environmental differences, it was still apparent that some environmental stimuli combinations were facilitating aphid development in some fir communities. Communities on southwestern exposures were heavily infested by aphids. The environment was favorable for the aphid and also the host by providing a conducive environment for extended growth.

There is no reason to suspect that stands growing on eastern aspects will not ultimately be colonized by aphids. Our results indicate that infestations begin their development on the southwestern exposures. They will, however, move to the eastern exposures with time and colonize the entire mountain.

CHAPTER VI

MANAGEMENT RECOMMENDATIONS

There are no known, effective techniques for controlling balsam woolly aphid populations on large areas anywhere in North America. Those that have shown promise focus on chemical insecticides, which are not compatible with wilderness areas of the National Park Service. If developing a system of aphid control was a major goal of this project, we have failed. Our research has not illuminated any technique to inhibit balsam woolly aphids not already known; rather we have gathered and interpreted information as per the stated goals (page 13). Our field exposure during the two years of the project, albeit quite pleasant at times, has convinced us that balsam woolly aphids are everywhere and their numbers are so overwhelming that cold weather, snow or rain, dry, hot temperatures, or a few predators will never dent the population significantly enough to stall the advancing hordes!

While a valuable resource is severely damaged, will it be possible to manage the situation to minimize the impact? These management recommendations were designed to assist National Park Service management personnel cope with the developing crisis. Unfortunately we do not know how to stop the impending color displays in the spruce-fir forests of the Great Smoky Mountains National Park.

113

Threat of Wildfire

Wildfire presents a number of problems in aphid-killed fir stands. Ecologically, wildfire does not have a meaningful role in natural spruce-fir stands. One aerial view of Mt. Sterling will confirm how wildfire affects spruce and fir. Therefore, fire protection can be within the intent of wilderness.

Wildfire poses a real threat to fir seedlings and saplings, regardless of the condition of the overstory. However, since the fir seed supply will soon be gone from the Park, wildfire in regenerating fir will cause an immediate species conversion. There are some yellow birch stands on Mt. Sterling that trace their origins to the early wildfires. Of course, once a wildfire develops intensity, it will not stop after exhausting the fir fuels.

Two management practices recommended in addition to the usual detection schemes employed are fire risk manipulation and hazard reduction. People cause most wildfires in the Park; those that originate from ignorance and carelessness are more easily controlled than those of malicious origin. Programs designed for decreased visits during peak fire danger periods in areas of high hazard, including closing the Dome Road or the Appalachian Trail, may be warranted. Existing trails through high hazard areas could be relocated in conjunction with environmental impact relocations.

Hazard reduction practices warrant consideration due to fire control, but they also warrant consideration due to understory

release. Soon after dying, the fir frame is festooned with epiphytic lichens, thereby creating extremely dangerous fuels. Not only do they ignite and burn rapidly, but they greatly increase resistance to control by spotting flaming embers for long distances. Snag-felling operations in critical areas will be good investments in wildfire prevention.

Community Manipulation

With mature fir holding tenaciously to a limited existence, regeneration of fir stands decimated by the aphid assume prime importance. There are a number of practices that can facilitate fir regeneration. In most cases, the obvious and first step is release of the fir understory. Such an operation also has hazard reduction benefits.

There is good silvicultural evidence that many species of Abies are capable of good growth after being released from overtopping trees. Partial release may be best in some cases, but total release works well if seedlings have developed roots in the mineral soil. Seedlings rooted in the humus layers after release will die during periods of moisture stress. Fraser fir responds well to release.

If stands carefully selected according to understory species composition and stage of development were released from all overstory trees, even-aged stands of Fraser fir would result. These dense, young stands were generally found to be aphid-free. Assuming the released trees could reproduce before being killed by balsam woolly

aphids, the fir could be perpetuated in natural stands. Otherwise the species might only exist in memories and museums.

There are several managerial practices that professional foresters use to increase seed production. Thinning Fraser fir stands that are approaching reproductive size and remain free of aphids has potential. Pole-sized, uninfested stands could be pushed toward seed production to secure advanced reproduction before the aphid kills the overstory. Thinning by herbicide, girdling, or felling the trees to free those showing good seed production potential is possible. Crown release and subsequent growth increase loci for cone production and the necessary energy for cone maturation.

Cooperative Seed Orchard

There is commercial importance to Fraser fir through landscaping and Christmas trees. As available seed sources diminish, more than a scenic resource will be lost.

It is our recommendation that the National Park Service pursue a cooperative program with the Forest Service and various states to build and maintain a seed orchard for the perpetuation of an endangered species. If the Service initiated the program and supplied seed or scionwood, cooperating agencies could supply materials and supporting personnel to implement the field work; it's worth investigating! Consider establishing the field site well beyond the probable distribution of the balsam woolly aphid. Upper Michigan or northern Minnesota are possibilities, although balsam woolly aphids

are still expanding their distribution in the East and West (Toko and Knauer, 1977).

Summary of Management Recommendations

1) Initiate and maintain fire prevention programs by minimizing visitors (fire risks) and reducing fuels (fire hazard) in high-use areas.

2) Manipulate stands to facilitate development of Fraser fir reproduction, either by releasing advanced reproduction or encouraging seedling formation in existing stands.

3) Initiate a cooperative program to insure perpetuation of an endangered species.

BIBLIOGRAPHY

BIBLIOGRAPHY

Aldrich, R. C. and A. T. Drooz. 1967. Estimated Fraser fir mor-
    tality and balsam woolly aphid infestation trends using
    aerial color photography. For. Sci. 13(3):300-313.

Amman, G. D. 1962. Seasonal biology of the balsam woolly aphid
    on Mt. Mitchell, North Carolina. J. Econ. Ent. 55(1):96-98.

Amman, G. D. 1966. Some new infestations of balsam woolly aphid
    in North Carolina, with possible modes of dispersal. J. Econ.
    Ent. 59(3):508-11.

Amman, G. D. 1967. Effect of -29° F. on over-wintering populations
    of the balsam woolly aphid in North Carolina. J. Econ. Ent.
    60(6):1765-1766.

Amman, G. D. 1968. Effects of temperature and humidity on devel-
    opment and hatching of eggs of Adelges piceae. Ann. Ent. Soc.
    Amer. 61(6):1606-1611.

Amman, G. D. 1969. A method of sampling balsam woolly aphid on
    Fraser fir in North Carolina. Can. Ent. 101(8):883-889.

Amman G. D. 1970. Distribution of redwood caused by balsam
    woolly aphid in Fraser fir of North Carolina. Res. Note
    SE-135, SE For. Exp. Sta. 4 p.

Amman, G. D. 1970. Phenomena of Adelges piceae populations
    (Homoptera:Phylloxeridae) in North Carolina. Ann. Ent.
    Soc. Amer. 63(6):1727-1734.

Amman, G. D. 1971. Infestation trends of balsam woolly aphid in
    an Abies alba plantation in North Carolina. Res. Note SE-148,
    SE For. Exp. Sta. 6 p.

Amman, G. D. and C. F. Speers. 1965. Balsam woolly aphid in
    the Southern Appalachians. J. For. 63(1):18-20.

119

Amman, G. D. and C. F. Speers. 1971. Introduction and evaluation of predators from India and Pakistan for control of the balsam woolly aphid. Can. Ent. 103:528-533.

Amman, G. D. and R. L. Talierico. 1967. Symptoms of infestation by balsam woolly aphid displayed by Fraser fir and bracted balsam fir. Res. Note SE-85, SE For. Exp. Sta. 7 p.

Annand, P. N. 1928. A contribution toward a monograph of the Adeliginae (Phylloxeridae) of North America. Stanford Univ. Press, Palo Alto, Calif. 146 p.

Atkins, M. D. 1972. Developmental variability among laboratory reared balsam woolly aphid (Hemiptera:Chermidae). Can. Ent. 104(2):203-208.

Atkins, M. D. and A. A. Hall. 1969. Effect of light and temperature on the activity of the balsam woolly aphid crawlers. Can. Ent. 101:481-488.

Balch, R. E. 1952. Studies of the balsam woolly aphid, Adelges picae (Ratz.) and its effects on balsam fir, Abies balsamea (L.) Mill. Canada Dept. of Agric. Publ. #867, 76 p., illustrated.

Balch, R. E., J. Clark and J. M. Bonga. 1964. Hormonal action in production of tumors and compression wood by an aphid. Nature. 202(4933):721-722.

Barr, A. J. and J. H. Goodnight. 1972. A user's guide to the statistical analysis system (SAS). Student Supply Stores, North Carolina State Univ., Raleigh, N. C. 260 p.

Barr, A. J., J. H. Goodnight, J. R. Sall and J. T. Helwig. 1976. The user's guide to SAS, 1976. Statistical Analysis Systems Institute, Inc. Raleigh, North Carolina. 260 p.

Bonga, J. M. and J. Clark. 1965. The effect of B-inhibitor on histogenesis of balsam fir bark cultured in vitro. For. Sci. 11(3):271-278.

Brower, A. E. 1947. Balsam woolly aphid in Maine. J. Econ. Ent. 40(5):689-694.

Bryant, D. G. 1963. Means of dispersal and recommended quarentine practices for balsam woolly aphid. Phytoprotection. 44:42-46.

120

Bryant, D. G. 1971. Balsam woolly aphid, Adelges piceae (Homoptera:Phylloxeridae), seasonal and spatial development on branches of balsam fir, Abies balsamea. Can. Ent. 103(10): 1411-1420.

Bryant, D. G. 1974. Distribution of first instar nymphs of Adelges piceae on branches of balsam fir after colonization. Can Ent. 106(10):1075-1080.

Bryant, D. G. 1974. A review of the taxonomy, biology and importance of the adelgid pests of true firs. Can. Environ., For. Serv., NFLD. For. Res. Centre. No. N-X-111. 50 p.

Bryant, D. G. 1976. Sampling populations of Adelges piceae on balsam fir, Abies balsamea. Can. Ent. 108:1113-1124.

Carroll, W. J. and D. G. Bryant. 1960. A review of balsam woolly aphid in Newfoundland. For. Chron. 36(3):278-290.

Carrow, J. R. 1969. An apparatus for field study of the balsam woolly aphid. Can. Ent. 101(2):132-134.

Carrow, J. R. and R. E. Betts. 1973. Effects of different foliar-applied nitrogen fertilizers on the balsam woolly aphid. Can. J. For. Res. 3(1):122-139.

Carrow, J. R. and K. Graham. 1968. Nitrogen fertilization of the host tree and population growth of the balsam woolly aphid. Can. Ent. 100(5):478-485.

Chew, V. 1976. Comparing treatment means: a compendium. Hort. Sci. 11(4):348-356.

Ciesla, W. M. and W. D. Buchanan. 1962. Biological evaluation of balsam woolly aphid, Roan Mtn. Gardens, Toecane District, Pisgah National Forest, North Carolina. USFS, SA. S & PF, Div. FPM. Report 62-93 (unpublished).

Ciesla, W. M., H. L. Lambert and R. T. Franklin. 1963. The status of the balsam woolly aphid in North Carolina and Tennessee. USFS, S & PF, Div. FPM, Ashville, North Carolina. Report No. 1-11-63 (unpublished).

Ciesla, W. M., H. L. Lambert, and R. T. Franklin. 1965. Status of the balsam woolly aphid - 1964. USFS, S & PF, Div. FPM, Ashville, North Carolina. Report No. 65-1-1.

121

Crandall, D. L. 1958. Ground vegetation patterns of the spruce-fir area of the Great Smoky Mountains National Park. Ecol. Mono. 28(4):337-360.

Doerksen, A. K. and R. G. Mitchell. 1965. Effects of balsam woolly aphid on the wood anatomy of some western true firs. For. Sci. 11(2):181-188.

Edwards, D. K. 1966. Observations on the crawlers of the balsam woolly aphid, Adelges piceae (Ratz.). Can. Dep. For. Rural Develop., Bi-Mon. Res. Notes, 22(5):4-5.

Fedde, G. F. 1973. Cone production in Fraser firs infested by the balsam woolly aphid, Adelges piceae (Homoptera:Phyllox-eridae). J. Georgia Ent. Soc. 8(2):127-131.

Fedde, D. G. 1973. Impact of the balsam woolly aphid (Homoptera: Phylloxeridae) on cones and seed produced by infested Fraser fir. Can. Ent. 105(5):673-680.

Fedde, G. F. 1974. A bark fungus for identifying Fraser fir irreversibly damaged by the balsam woolly aphid. J. Georgia Ent. Soc. 9(1):64-68.

Flavell, T. H. and H. L. Lambert. 1971. Status of the balsam woolly aphid in the Southern Appalachians - 1970. USFS, SA, S & PF, Div. FPM, Ashville, North Carolina. Report No. 71-1-16.

Forbes, A. R. and D. B. Mullick. 1970. The stylets of the balsam woolly aphid. Can. Ent. 102(9):1074-1082.

Foulger, A. N. 1968. Effect of aphid infestations on the properties of grand fir. For. Prod. J. 18(1):43-47.

Greenbank, D. O. 1970. Climate and ecology of the balsam woolly aphid. Can. Ent. 102(5):546-578.

Hopewell, W. W. and D. G. Bryant. 1969. Chemical control of Adelges piceae (Homoptera:Adelgidae) in Newfoundland, 1967. Can. Ent. 101:1112-1114.

Hudak, J. and R. E. Wells. 1974. Armillaria root rot in aphid-damaged balsam fir in Newfoundland. For. Chron. 50(2):74-76.

122

Johnson, N. E., R. G. Mitchell and K. H. Wright. 1963. Mortality and damage to Pacific silver fir by the balsam woolly aphid in southwestern Washington. J. For. 61(11): 854-860.

Johnson, N. E. and J. G. Zingg. 1968. The balsam woolly aphid on young silver fir in Washington. Weyerhaeuser For. Pap., Centralia, Washington. No. 13.

Kotinsky, J. 1916. The European fir trunk louse, Chermes (Dreyfusia) piceae (Ratz.). Ent. Proc. Soc. Washington. 18:14-16.

Lambert, H. L. and W. M. Ciesla. 1966. Status of the balsam woolly aphid in North Carolina and Tennessee - 1965. USFS, SA, S & PF, Div. FPM, Ashville, North Carolina. Report No. 66-1-1.

Lambert, H. L. and W. M. Ciesla. 1967. Impact of summer cutting on the dispersal of the balsam woolly aphid. J. Econ. Ent. 60(2):613-614.

Lambert, H. L. and W. M. Ciesla. 1967. Status of the balsam woolly aphid in the Southern Appalachians - 1966. USFS, SA, S & PF, Div. FPM, Ashville, North Carolina. Report No. 67-1-3.

Lambert, H. L. and R. T. Franklin. 1967. Tanglefoot traps for detection of the balsam woolly aphid. J. Econ. Ent. 60(6): 1525-1529.

Lambert, H. L. and J. L. Rauschenberger. 1968. Status of the balsam woolly aphid in the Southern Appalachians - 1967. USFS, SA, S & PF, Div. FPM, Ashville, North Carolina. Report No. 68-1-17.

McCambridge, W. F. 1958. Detection and appraisal survey of the balsam woolly aphid on Mt. Mitchell State Park and the North Carolina National Forest. SE For. Expir. Sta., Ashville, North Carolina. Report No. 58-5 (unpublished).

McMullen, L. H. and J. P. Skovsgaard. 1972. Seasonal history of the balsam woolly aphid in coastal British Columbia. J. Ent. Soc. B. C. 69:33-40.

Mitchell, R. G. 1966. Infestation characteristics of the balsam woolly aphid in the Pacific Northwest. USFS Res. Pap. PNW-35, 18 p.

123

Mitchell, R. G., G. D. Amman and W. E. Waters. 1970. Balsam
    woolly aphid. USFS, For. Pest Leaflet 118. 10 p.

Mitchell, R. G. M. E. Johnson and S. A. Rudinsky. 1961. Seasonal
    history of the balsam woolly aphid in the Pacific Northwest.
    Can. Ent. 93(9):794-798.

Mitchell, R. G. and K. H. Wright. 1967. Foreign predator intro-
    duction for control of the balsam woolly aphid in the Pacific
    Northwest. J. Econ. Ent. 60(1):140-147.

Nagel, W. P. 1959. Forest insect conditions in the Southeast
    during 1958. USFS, Sta. Paper SE-100. 10 p.

Oosting, H. J. and W. D. Billings. 1951. A comparison of virgin
    spruce-fir forest in the Northern and Southern Appalachian
    system. Ecol. 42:84-103.

Page, G. 1975. The impact of balsam woolly aphid damage on
    balsam fir stands in Newfoundland. Can. J. For. Res.
    5(2):195-209.

Purtich, G. S. 1971. Water permeability of the wood of grand
    fir (Abies grandis (Doug.) Lindl.) in relation to infestation
    by the balsam woolly aphid, Adelges Piceae (Ratz.). J. Exptl.
    Bot. 22:936-945.

Purtich, G. S. 1973. Effect of water stress on photosynthesis,
    respiration and transpiration of four Abies species. Can.
    J. For. Res. 3:293-298.

Purtich, G. S. 1977. Distribution and phenolic composition of
    sapwood and heartwood in Abies grandis (Doug.) Lindl. and
    effects of the balsam woolly aphid. Can. J. For. Res.
    7(1):54-62.

Purtich, G. S. and R. P. C. Johnson. 1971. Effects of infesta-
    tion by the balsam woolly aphid, Adelges piceae (Ratz.), on
    the ultrastructure of bordered-pit membranes of grand fir,
    Abies grandis (Day.) Lindl. J. Exptl. Bot. 22:953-958.

Purtich, G. S. and W. W. Nijholt. 1974. Occurrence of juvabione-
    related compounds in grand fir and Pacific silver fir in-
    fested by the balsam woolly aphid. Can. J. Bot. 52(3):
    585-587.

124

Puritch, G. S. and J. A. Petty. 1971. Effect of balsam woolly aphid, Adelges piceae (Ratz.), infestation on the xylem of Abies grandis (Doug.) Lindl. J. Exptl. Bot. 22:946-952.

Rauschenberger, J. L. and H. L. Lambert. 1970. Status of the balsam woolly aphid in the Southern Appalachians - 1969. USFS, SA, S & PF, Div. FPM, Ashville, North Carolina. Report No. 70-1-44.

Reed, D. 1964. Fraser fir - Vanishing species? Ashville Citizen Times, Oct. 25, 1964.

Schooley, H. O. 1976. Recovery of young balsam fir trees damaged by the balsam woolly aphid. For. Chron. 52:143-144.

Schooley, H. O. and L. Oldford. 1974. Balsam woolly aphid damage to the crowns of balsam fir trees. Info. Rpt., NFLD. For. Res. Cen. No. N-X-121, 27 p.

Schooley, H. O. and L. Oldford. 1974. Damage caused by the balsam woolly aphid in young balsam fir stands. Info. Rpt. NFLD. For. Res. Cen. No. N-X-115, 19 p.

Smith, B. C. 1958. Responses to light and influence of light and temperature on locomotion of crawlers of the balsam woolly aphid and of insect predators of this species. Can. Ent. 90(4):193-201.

Speers, C. F. 1958. The balsam woolly aphid in the Southeast. J. For. 56:515-516.

Thor, E. and P. Barnett. 1974. Taxonomy of Abies in the Southern Appalachians: variations in balsam monoterpenes and wood properties. For. Sci. 20(1):32-40.

Toko, H. V. and K. H. Knauer. 1977. Forest insect and disease conditions in the United States - 1975. U. S. Government Printing Office, No. 0-226-535, 60 p.

Vyse, A. H. 1971. Balsam woolly aphid; a potential threat to the British Columbia forests. Dept. Environ., Can. For. Serv., Pac. For. Res. Centre, Info. Report BC-X-61.

Ward, J. D., E. T. Wilson and W. M. McDonell. 1973. Status of the balsam woolly aphid, Adelges piceae (Ratz.), in the Southern Appalachians - 1972. USFS, SA, S & PF, Div. FPM, Ashville, North Carolina. Report No. 73-1-35.

125

Whittaker, R. H. 1956. Vegetation of the Great Smoky Mountains. Ecol. Monographs. 26(1):1-80.

Wood, R. O. 1968. First occurrence of the balsam woolly aphid in interior British Columbia. J. Ent. Soc. Brit. Col. 65:13-14.

Woods, T. A. and M. D. Atkins. 1967. A study of the dispersal of balsam woolly aphid crawlers by small animals. Can. Dep. For. and Rural Develop. Bi-Mon. Res. Notes 23(6):44.

CPSIA information can be obtained
at www.ICGtesting.com
Printed in the USA
BVHW04*1008190918

527934BV00014B/692/P

9 780332 730097